T0211337

A Method to Identify Energy Efficiency Measures for Factory Systems Based on Qualitative Modeling

Manuela Krones

A Method to Identify Energy Efficiency Measures for Factory Systems Based on Qualitative Modeling

 Springer Vieweg

Manuela Krones
Chemnitz, Germany

Diese Arbeit wurde von der Fakultät für Maschinenbau der Technischen Universität Chemnitz als Dissertation zur Erlangung des akademischen Grades Doktoringenieur (Dr.-Ing.) genehmigt.

Tag der Einreichung: 19. Oktober 2016
Betreuer: Prof. Dr.-Ing. Egon Müller
1. Gutachter: Prof. Dr.-Ing. Egon Müller
2. Gutachter: Frank E. Pfefferkorn, Ph.D.
Tag der Verteidigung: 13. Februar 2017

ISBN 978-3-658-18342-4   ISBN 978-3-658-18343-1   (eBook)
DOI 10.1007/978-3-658-18343-1

Library of Congress Control Number: 2017940984

Springer Vieweg
© Springer Fachmedien Wiesbaden GmbH 2017

This Springer Vieweg imprint is published by Springer Nature
The registered company is Springer Fachmedien Wiesbaden GmbH
The registered company address is: Abraham-Lincoln-Str. 46, 65189 Wiesbaden, Germany

Preface

Due to changing conditions in politics and laws, energy-saving products and technologies get increasingly important. A major future challenge is to generate economic growth while reducing the input of resources. Besides economic objectives, energy efficiency is currently getting significant for the competitiveness of industrial enterprises. Therefore, professional and methodical competencies for planning and operating energy-efficient factories are required. The field of factory planning bears a special responsibility since decisions during early planning phases mainly influence the relevant characteristics of factories in terms of energy efficiency.

Systematically considering energy efficiency as an objective during factory planning requires suitable methods and tools that support planning participants in identifying improvement opportunities. This topic is addressed by the author of this dissertation thesis, Ms. Manuela Krones. The focus of the thesis is to develop and apply a method for identifying energy efficiency measures for factory systems based on qualitative modeling. The developed method provides qualitative description concepts for factory planning tasks and energy efficiency knowledge as well as an algorithm-based linkage between these measures and the respective planning tasks. The application of the method is guided by a procedure model, which allows a general applicability in the manufacturing sector. It should be emphasized that the object area of factory systems is explained systematically and on a high level of detail, which is framed by comprehensible examples. The validation by means of two case studies demonstrates the applicability of the method in various planning areas. The method leads to the desired results in terms of energy efficiency improvements while the effort for the application is reduced. Furthermore, the case studies highlight the importance of cooperation between planning participants in order to increase energy efficiency.

Ms. Krones has excelled in describing relevant aspects clearly and understandingly to the reader. She systematically analyzes and describes theoretical approaches, suitable description concepts, and mathematical contexts. Based on that, the methodical contribution is developed and explained in detail with a close reference to the practical usage.

This thesis provides a significant impact on research and practice in the area of factory planning with regard to energy efficiency and gives an excellent support for scientists and practitioners in this field.

Prof. Dr.-Ing. Egon Müller

Foreword

Die vorliegende Arbeit entstand während meiner Tätigkeit als wissenschaftliche Mitarbeiterin an der Professur Fabrikplanung und Fabrikbetrieb der Technischen Universität Chemnitz. Während dieser Zeit habe ich von vielen Personen Unterstützung erhalten, denen ich hiermit meinen Dank aussprechen möchte.

Ganz besonders danken möchte ich meinem Doktorvater Prof. Dr.-Ing. Egon Müller, Leiter des Institutes für Betriebswissenschaften und Fabriksysteme und Inhaber der Professur Fabrikplanung und Fabrikbetrieb, für die Betreuung und wohlwollende Förderung sowie den gewährten inhaltlichen Freiraum.

During my research work, I spent a few months at the Laser-Assisted Multi Scale Manufacturing Laboratory of the University of Wisconsin-Madison, USA. A special thank goes to Frank E. Pfefferkorn, Ph.D. for hosting my research stay and supervising my dissertation thesis as second reviewer. Thanks to the entire team for making the research stay a rewarding experience and for the final proofreading of my thesis. Furthermore, I'd like to thank the German Academic Exchange Service (DAAD) for the funding.

Weiterhin möchte ich Herrn Prof. Dr. Uwe Götze, Inhaber der Professur Unternehmensrechnung und Controlling, für die langjährige Unterstützung danken, die ich bereits während meines Studiums erfahren durfte.

Großer Dank gilt allen (ehemaligen) Kollegen des Institutes für Betriebswissenschaften und Fabriksysteme für die sehr gute Zusammenarbeit und das kooperative und freundschaftliche Arbeitsklima. Ganz besonders danke ich Herrn Dr.-Ing. Hendrik Hopf für die zahlreichen inhaltlichen Anregungen und Herrn Dipl.-Ing. Frank Börner für die Unterstützung bei der Gestaltung der Arbeit. Meinen Mitdoktoranden Dipl.-Wirt.-Ing. (FH) Martin Domagk, Dr.-Ing. David Jentsch, Dipl.-Wi.-Ing. Andreas Merkel, Dr.-Ing. Daniel Oehme, Dr.-Ing. Daniel Plorin und Dr.-Ing. Timo Stock danke ich für den fortwährenden Austausch und die gegenseitige Motivation. Bei Herrn Dipl.-Ing. Gert Kobylka bedanke ich mich für die angenehme Zusammenarbeit bei allen verwaltungstechnischen Belangen und für die Korrektur der deutschsprachigen Unterlagen.

Meinen Freunden möchte ich dafür danken, dass sie mich stets motiviert und bei Bedarf für die notwendige Zerstreuung gesorgt haben. Herrn Dr. rer. nat. Heiko Weichelt danke ich für die hilfreichen Anmerkungen zu Rechtschreibung und Grammatik.

Besonders herzlich bedanken möchte ich mich bei meiner Familie für die uneingeschränkte Unterstützung, das entgegengebrachte Verständnis und den gebotenen Rückhalt.

Chemnitz, Februar 2017   Manuela Krones

Abstract

Energy efficiency is an important objective for industrial enterprises. Since the energy consumption of a factory is mainly influenced during early planning phases, there is a need for methods and tools to support the integration of the objective energy efficiency into factory planning. The goal of this thesis is to develop a method to identify energy efficiency measures for factory systems which can be applied preferably during factory planning. While existing methods for energy efficiency improvement usually contain a quantitative assessment, this work suggests the identification of measures based on qualitative models of the factory system. The main support is to generate solution approaches for increasing energy efficiency. The research contribution lies in a method that contains qualitative description concepts for both factory planning tasks and energy efficiency knowledge, an algorithm to suitably assign energy efficiency measures to factory planning tasks, and a procedure model for the method's application.

Contents

List of Figures

List of Tables

List of Abbreviations

AGV	Automated Guided Vehicle
ARIS	Architecture of Integrated Information Systems
AWS	American Welding Society
BAT	Best Available Technique
BDEW	German Association of Energy and Water Industries (Bundesverband der Energie- und Wasserwirtschaft e. V.)
BKCASE	Body of Knowledge and Curriculum to Advance Systems Engineering
BMWi	German Federal Ministry for Economic Affairs and Energy (Bundesministerium für Wirtschaft und Energie)
BPMN	Business Process Model and Notation
dena	German Energy Agency (Deutsche Energieagentur)
DMM	Domain Mapping Matrix
DSM	Design Structure Matrix
EDL-G	Energy Service Law (Energiedienstleistungsgesetz)
EEG	Renewable Energy Sources Act (Erneuerbare-Energien-Gesetz)
EEM	Energy Efficiency Measure
EEMA	Energy Efficiency Measure Matching Algorithm
EEMIS	Energy Efficiency Measure Implementation Support
EEWaermeG	Renewable Energy Heat Law (Erneuerbare-Energien-Wärmegesetz)
EnEV	Energy Saving Directive (Energieeinsparverordnung)
EVPG	Energy-Related Products Law (Energieverbrauchsrelevante-Produkte-Gesetz)
EVS	Energy Value Stream
FSW	Friction Stir Welding
GMAW	Gas-Metal Arc Welding
GTAW	Gas-Tungsten Arc Welding
HOAI	Official Scale of Fees for Services by Architects and Engineers (Honorarordnung für Architekten und Ingenieure)
IEA	International Energy Agency
IEEE	Institute of Electrical and Electronics Engineers
IPCC	Intergovernmental Panel on Climate Change
LHAW	Laser-Hybrid Arc Welding

MAG	Metal Active Gas Welding
MDM	Multiple-Domain Matrix
MIG	Metal Inert Gas Welding
MMAW	Manual Metal-Arc Welding
NAICS	North American Industry Classification System
PDCA	Plan-Do-Check-Act Cycle
SADT	Structured Analysis and Design Technique
SMAW	Submerged Arc Welding
SME	Small and medium-sized enterprise
StromStG	Energy and Electricity Tax Law (Stromsteuergesetz)
TAG	Tungsten Active Gas Welding
TIG	Tungsten Inert Gas Welding
TRIZ	Theory of Inventive Problem Solving
UML	Unified Modeling Language
vbw	Bavarian Industry Association (Verein der bayerischen Wirtschaft e. V.)

1 Introduction

The introduction briefly outlines the motivation for the topic and describes the research objectives. Afterwards, the research design and the structure of the thesis are explained.

1.1 Motivation

As a central challenge for the 21st century, the sustainable use of resources becomes an important objective. This development is driven by increasing worldwide greenhouse gas emissions, rising energy costs, scarcity of natural resources, and insecurities in the supply of energy and resources. The increasing anthropogenic greenhouse gas emissions lead to climate changes, such as global warming (Intergovernmental Panel on Climate Change, IPCC, 2015, p. 4).

The concept of sustainability has its origin around 1700 in the context of forestry: it meant to cut only as much trees as can grow in the same period of time (von Carlowitz, 2013). The modern understanding of sustainability is mainly characterized by the World Commission on Environment and Development which defines *sustainable development* as a "development that meets the needs of the present without compromising the ability of future generations to meet their own needs" (World Commission on Environment and Development, 1987, p. 41). This means that, despite economic growth, the availability of resources should be preserved. In the common understanding, sustainability can be realized by considering economic, ecological, and social objectives as equivalent in the so-called triple bottom line (Elkington, 1997).

Industrial enterprises have a relevant share of the global energy consumption and green-house gas emissions. In this context, sustainable manufacturing is understood as "creation of goods and services using processes and systems that are non-polluting, conserving of energy and natural resources, economically viable, safe, and healthful for workers, communities and consumers, and socially and creatively rewarding for all working people" (The Lowell Center for Sustainable Production, 2016). This means that the planning and operation of factories need to consider related economic, ecological, and social effects.

Within the general area of environmentally sustainable manufacturing systems, the design of energy-efficient factories has major relevance (Haapala et al., 2013, p. 9; Schenk, Wirth & Müller, 2014, p. 13). Factory planning is particularly important to realize the sustainability goals since the energy consumption of a factory is mainly influenced during early planning phases (Engelmann, Strauch & Müller, 2008, p. 61).

As a result, there is a need for methods and tools to identify energy efficiency potentials for factory systems. These instruments should support factory planning participants, which may come from various disciplines, and need to consider the high system complexity that is inherent to factory planning tasks.

1.2 Objectives

The main goal of the thesis is to develop a method that supports the increase of energy efficiency in factories. The focused application area is factory planning, although factory management tasks may be supported as well. Usually, only limited data and information on the energy consumption of production systems is available during early planning phases. Therefore, the method should be based on qualitative information, which is generally available in factory planning projects. Moreover, the method focuses on the optimization rather than on the analysis or assessment towards energy efficiency. This means that the desired main result after applying the method are suitable energy efficiency measures. Based on this goal, the thesis is guided by the following research questions:

- Which barriers do enterprises face towards energy efficiency improvements? How can these be overcome by a systematic procedure?

- How do existing procedures fulfill these requirements?

- How can qualitative information and knowledge be structured and represented in general? Which implications exist for factory planning?

- Which general procedure may be applied to identify energy efficiency potentials without quantitative information on energy consumption?

- Which objects in a factory system need to be considered for energy efficiency improvements?

- How are energy efficiency measures connected to the tasks of factory planning participants?

- What kind of energy efficiency knowledge needs to be provided by the methodical approach?

The development of the method needs to consider the requirements that arise from the practical application background. This means that the current barriers of enterprises to implement energy efficiency strategies and instruments are analyzed first. Based on this, a framework for the methodical procedure needs to be developed. This requires the application of methods and tools to represent qualitative information with regard to factory systems.

The method helps planning participants to identify energy efficiency measures for factory systems. This identification is based on qualitative models of the factory system. The procedure is structured as a step-by-step approach, whereof steps may be revised iteratively when requirements are changed or additional information becomes available. The purpose is to quickly generate solution approaches to increase energy efficiency. Hence, a concept to assign energy efficiency measures to factory planning tasks is required.

1.3 Research Design

The thesis addresses the area of factory planning as part of the engineering sciences, which are characterized as applied science. In contrast to that, formal sciences, such as mathematics and logics, deal with abstract structures rather than real, observable systems (Weber, Kabst & Baum, 2014, p. 25). While fundamental research focuses on the development of theoretical concepts, applied research focuses on practical problems or theoretical issues that are closely related to practice (Novikov & Novikov, 2013, p. 61). Fundamental research and applied research differ in terms of purpose, context and methods; *i.e.*, fundamental research intends to expand knowledge, while applied research aims at creating understanding of a specific problem (Hedrick, Bickman & Rog, 1993, pp. 2 ff.). Moreover, applied research develops guidelines for scientifically well-founded actions in a practical context (Ulrich, 1984, p. 171).

Defining the research design means to plan the research including the selection of methods and settings (Karlsson, 2009, p. 60). The research concept for this thesis is adapted from the strategy of applied research according to ULRICH (Ulrich, 1984, pp. 179 ff.). Applied research is characterized by using multiple research methods (Hedrick, Bickman & Rog, 1993, p. 9). This means that both theoretical and empirical research methods should be applied (Ulrich, 1984, p. 194). The research design and instruments for this thesis are depicted in Figure 1.

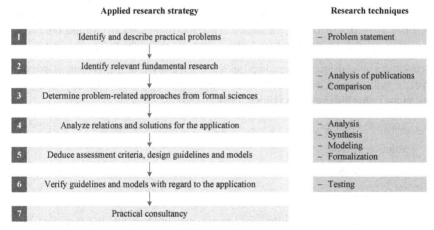

Figure 1: Research design of the thesis including applied research strategy (Ulrich, 1984, p. 193) and research techniques (Novikov & Novikov, 2013, pp. 44 ff.).

The first step of the research process is to identify problems in a practical context. This means that applied research questions arise from practical problems, for which a scientific solution (or solution procedure) is not yet available (Ulrich, 1984, p. 172). This thesis focuses on a method to identify energy efficiency measures for factory systems, hence, it relates to a strong practical background. The method *problem statement* means stating the research questions and constructing the expected result (Novikov & Novikov, 2013, p. 63). The research questions are initially described in Section 1.2; more detailed requirements are deduced based on the analysis of the state of the art.

Secondly, theories and hypotheses from fundamental sciences are identified and analyzed with regard to the specific problem. This means to identify relevant theoretical knowledge from various disciplines (Ulrich, 1984, p. 192). For this thesis, the fundamental basics mainly relate to the description of factory systems and factory planning tasks. The third step of the procedure includes the analysis of problem-relevant approaches from formal sciences. Since the purpose of the thesis is to provide information and knowledge to factory planning participants, important background from formal sciences is provided by methods and tools of knowledge management. Additionally, modeling and systems engineering are relevant fundamentals to describe objects and procedures.

The *analysis of publications* is a substantial part of any research process (Novikov & Novikov, 2013, p. 81). On the one hand, publications contain existing scientific ideas and concepts that need to be critically discussed before developing an own research contribution. On the other hand, documented scientific knowledge forms the methodological basis for the conducted research. Hence, present approaches to increase energy efficiency in factories are analyzed. A *comparison* is the judgment of similarities and differences between objects in order to identify characteristics or classify the objects (Novikov & Novikov, 2013, p. 45). Thus, a comparison between existing research contributions helps to identify relevant concepts that can be integrated into the method's development.

The fourth step of the procedure comprises a detailed analysis of the problem. Based on that, the main research result is developed in step five. The research result may belong to one of the following categories (Ulrich, 1984, p. 180):

- concrete solutions,

- solution methods,

- design models, or

- modeling guidelines.

A research result may be a solution for a concrete practical problem. In this context, the researcher is considered as consultant or expert in a specific area and is supposed to solve a problem. The second case does not refer to the direct solution of a problem but to the development of approaches for solving practical problems. A design model is understood as a model of a possible future scenario, which is developed by the researcher and put into application by a practitioner. Finally, the research result may be a guideline, which is used by practitioners to develop models. A major difference depends on whether a solution is developed for a product or an approach to achieve the solution (Ulrich, 1984, p. 180). In this thesis, a method is developed, which serves as a solution approach to solve practical problems.

The research techniques applied for the fourth and fifth step are the following: *Analysis* means to examine an object in detail and to identify its structure and characteristics (Novikov & Novikov, 2013, pp. 44 f.). In contrast to that, *synthesis* activities integrate different elements into a whole system in order to generalize knowledge (Novikov & Novikov, 2013, p. 45). Both activities may be used as an iterative, opposing process. *Modeling* is the process to generate an abstract representation from reality for further analyses or experiments. In this thesis, qualitative models are used to characterize factory planning tasks and are created through analysis and synthesis activities.

Formalization means to formally represent conceptual knowledge, *i.e.*, a logical or mathematical representation of information (Novikov & Novikov, 2013, p. 46). This is important for the assignment between energy efficiency measures and factory planning tasks.

After the development of guidelines and models, these are tested in a practical context as part of the sixth step. This test focuses on the applicability of a method or model, *i.e.*, whether the research result achieves a practical benefit (Ulrich, 1984, p. 175). Testing is an empirical method that may be carried out in different forms, such as questionnaires or practical works (Novikov & Novikov, 2013, p. 53). Since tests have a diagnostic purpose, they can be used to verify or validate the results of the research. Herein, a validation of the method is conducted. Finally, the developed results are to be applied in practical context, which, in general, is the main goal of applied research.

1.4 Structure of the Thesis

The structure of the thesis is deduced from the research strategy. It includes the main steps to analyze the state of the art from both a practical and scientific point of view. Afterwards, requirements for the method are deduced; based on these, the method and its components are developed and tested.

An overview of the chapters and their main content is depicted in Figure 2. As part of this first chapter, the motivation towards the topic is presented and research objectives are deduced. Based on these goals, a corresponding research design is developed.

Chapter 2 addresses the energy use in manufacturing industry. It analyzes drivers that promote the need to increase energy efficiency from an economic, ecological, and political perspective. Moreover, barriers towards the implementation of energy efficiency strategies are analyzed in order to consider these as requirements for the method's development.

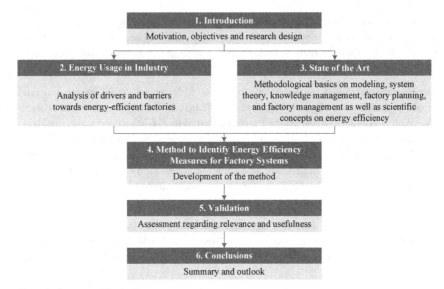

Figure 2: Structure of the thesis

The third chapter analyzes the scientific state of the art. First, basics on models and modeling are presented as the basis for the qualitative modeling in the thesis (Section 3.1). Then, system theory and systems engineering are explained – both for the description of factory systems and as a technique for problem-solving (Section 3.2). Afterwards, basics on knowledge management are described since the purpose of the method is to provide relevant information and knowledge to factory planning participants (Section 3.3). An important basis for the formalization of concepts is given by the techniques of knowledge representation. In Section 3.4, factory planning tasks and approaches are presented. Moreover, an understanding for factories as socio-technical systems is generated. Finally, existing methods and tools to support the energy efficiency optimization of factories are described and evaluated in Section 3.5.

The main part of the thesis is the development of the method in Chapter 4. Based on the findings from the previous chapters, requirements for the method are formulated (Section 4.1). Subsequently, the framework for the method is deduced and the relevant components of the research contribution are presented (Section 4.2). Afterwards, description concepts for each of the identified domains, which cover both the factory planning task and the energy efficiency knowledge, are discussed in detail (Sections 4.3 to 4.8). Then, the algorithm to match energy efficiency measures to factory planning tasks is developed in Section 4.9. Finally, the previous insights are integrated into a procedure model for the method's application in Section 4.10.

The fifth chapter addresses the validation of the method with regards to the criteria of relevance and usefulness. The validation is conducted by assessing the methodical requirements, implementing a prototype and conducting two comprehensive case studies. Eventually, the sixth chapter gives a summary on the thesis and an outlook on future research work.

2 Energy Usage in Industry

The second chapter illustrates fundamental concepts and underlines the motivation for the research work. At first, relevant terms are explained. Afterwards, drivers and barriers for increasing energy efficiency in industrial enterprises are analyzed with regard to economic, ecological, and political factors. The identification of barriers that hinder the implementation of energy efficiency strategies form a starting point to define requirements for the method.

2.1 Terms and Definitions

Energy E is a state variable of a system that describes its ability to perform work. If a system passes from one state to another one, *work W* appears as a variable to describe this process. Work, heat, and energy are measured by the unit Joule [*J*], *i.e.*, using the SI units:

$$1 \ J = 1 \ N \cdot m = 1 \ W \cdot s = 1 \ \frac{kg \cdot m^2}{s^2}. \tag{2.1}$$

The work of a process related to one period of time is the *power P* measured in Watt [*W*].

The laws of thermodynamics determine fundamental rules on the transformation of energy. The first law of thermodynamics describes the conservation of energy. It states that the total energy of an isolated system is constant, *i.e.*, energy cannot be created or destroyed. However, energy may be transformed from one form into another one. Energy is composed of exergy, which is usable to perform work, and anergy, which may not be used within the existing surrounding. Although the energy within a system remains constant, it converts exergy into anergy, which is often referred to as energy loss or energy consumption. This direction implies the irreversibility of a process, which is stated by the second law of thermodynamics. Accordingly, heat can never pass from a colder to a warmer body without other changes. Therefore, natural processes take place in a certain direction and are not reversible (*e.g.*, friction). (Windisch, 2014, pp. 70 ff.)

Energy occurs in various *energy forms*, such as mechanical (potential and kinetic), chemical, electrical, thermal, and nuclear (Hesselbach, 2012, p. 18). When considering the physical objects that hold this energy, several *energy carriers* may be identified, such as gas (*e.g.*, natural gas, steam, compressed air), liquids (*e.g.*, oil), and solid materials (*e.g.*, biomass, coal, uranium).

Transformation processes are necessary to make energy carriers useful for any application. The energy conversion chain contains the transformation from raw material until the final application. Energy losses occur on every conversion step. The major phases within this transformation process are explained by the following terms (Müller et al., 2009, pp. 72 ff.): *Primary energy* is the energy content of natural resources prior to any transformation (*e.g.*, natural gas, wind). After any transformation, the energy is referred to as *secondary energy* (*e.g.*, diesel). The *end energy* is provided to the final user, for example to a production plant (*e.g.*, electricity). Finally, the energy is directly applied for a specific purpose which is called *use energy* (*e.g.*, lighting).

The objective to increase energy efficiency needs to be discussed with regard to the general ideas of efficiency, effectiveness, and productivity: *Productivity* describes the ratio between an output, such as the amount of products, and the input of production factors, such as time, material, or staff (Nebl, 2011, pp. 18 f.). *Efficiency* is the ratio between useful output and input, whereas *effectiveness* means the ability to produce a desired output (Miller, Colombi & Tvaryanas, 2014, p. 205). A demonstrative definition of the latter two terms is given with efficiency as "doing things right" and effectiveness as "doing the right things" (Drucker, 1974, p. 83).

Energy efficiency can be defined as the ratio between a useful output and the input of energy that is necessary to achieve this output (Müller et al., 2009, p. 2):

$$\text{energy efficiency} = \frac{\text{useful output}}{\text{energy input}}. \tag{2.2}$$

Thus, increasing energy efficiency can be achieved with two strategies: The output can be increased while maintaining a constant energy consumption or the energy consumption needs to be reduced while maintaining the useful output. The usual approach to increase energy efficiency, which is also pursued in this thesis, is to reduce the energy input. Energy effectiveness can hardly be expressed in a quantitative way, but it is used as a concept to question whether the energy provides a useful output. For example, air ventilation in a factory during idle time is an output that would not be considered useful.

Energy productivity is a less common term that is similar to energy efficiency. It expresses the useful output of the energy efficiency definition by means of economic objectives, for example the gross domestic product (Statistisches Bundesamt, 2014, p. 6). Another interpretation is the ratio between value added and energy costs which makes the energy productivity a percentage value (Reinhart et al., 2010, p. 870). The reciprocal of efficiency or productivity objectives represents the energy intensity (Linke et al., 2013, p. 557).

2.2 Driving Concerns for Energy Efficiency

A variety of external factors lead to the necessity to increase energy efficiency for industrial enterprises, such as economic, ecological, and political aspects. As a result, energy efficiency is considered as a relevant competitive factor (Bunse, Vodicka & Schönsleben, 2011, p. 53). A survey by the Fraunhofer Institute for Production Systems and Design Technology among 2,200 industrial companies, points out that 56 % of the participants confirm the importance of energy efficiency (Karcher & Siemer, 2013, p. 17).

When it comes to future development, even more than two third of the respondents expect a growing importance (Karcher & Siemer, 2013, p. 17). Main drivers that foster this development are explained in the following sections.

2.2.1 Ecological Effects of Energy Consumption

The emission of greenhouse gases that is caused by energy consumption considerably influences the global ecological system. Greenhouse gases absorb and emit infrared radiation and, thereby, contribute to global warming. The most important greenhouse gases are carbon dioxide, methane, nitrous oxides, and fluorinated gases (United States Environmental Protection Agency, 2016a).

Usually, ecological effects of these gases are expressed relatively to carbon dioxide (CO_2) since it has the highest share of greenhouse gas emissions. Hence, emissions of any other gas are expressed by carbon dioxide equivalents (CO_2e).

Industry contributes 20 % of the global carbon dioxide emissions and is responsible for an additionally 18 % due to allocated electricity and heat generation, see Figure 3 (International Energy Agency, IEA, 2014, p. 10). Therefore, reducing energy consumption is important in order to minimize global warming.

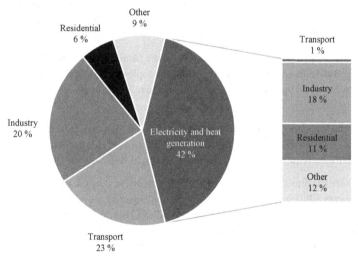

Figure 3: Global carbon dioxide emissions by sector in 2012 (International Energy Agency, IEA, 2014, p. 10)

The development of the global energy demand is decomposed into the following effects (International Energy Agency, IEA, 2012, p. 283): The *efficiency effects* mean to use less energy to provide the same level of service (*e.g.*, energy management in a company). The *fuel and technology switching effects* mean to provide a service with different fuels and technologies (*e.g.*, using natural gas instead of oil for room heating in buildings). The *activity effects* influence the demand for energy services (*e.g.*, economic growth).

According to the International Energy Agency (IEA), energy efficiency provides the highest single share of energy savings of these effects; it accounts for about 70 % in the future projection of the energy demand until 2035 (International Energy Agency, IEA, 2012, p. 282).

2.2.2 Political Conditions for Energy Efficiency

Political and legal conditions are the second group of drivers for industrial enterprises. These contain political strategies, laws, and governmental aids (*e.g.*, through providing financial aids for energy-efficient technologies). An overview with examples within these categories is given in the following paragraphs.

The European Commission concluded the Climate and Energy Framework in 2014 in order to set a European energy efficiency strategy (European Commission, 2016).[1] The key objectives of this framework target greenhouse gas emissions, renewable energy, and energy efficiency. According to the strategy, an increase in energy efficiency by at least 27 % by 2030 is aspired to (European Commission, 2016). A long-term strategy is defined by the Energy Roadmap 2050, which includes an 80-95 % reduction of greenhouse gas emissions by 2050 (European Union, 2012, p. 3).

The German Federal Government decided upon the *National Sustainability Strategy* in 2002 (Die Bundesregierung, 2002). Regarding the topic of energy efficiency, it contains the goal to double energy productivity by 2020 and to reduce primary energy consumption by 20 % by 2020.

The national legislation is influenced by directives and regulations from the European Union (EU). When a directive is passed from the European Commission, the national governments are required to implement these directives into national law.

Examples for EU directives in the context of energy efficiency are the Ecodesign Directive and the Energy Efficiency Directive. The *Ecodesign Directive* provides minimum requirements on the environmental performance of products (European Commission, 2009b), which are implemented through product-specific regulations, such as on the efficiency of electric motors (European Commission, 2014a). The *Energy Efficiency Directive* requires the national governments to define national strategies on energy efficiency and to report their achievements (European Parliament, 2012).

---

[1] Before, the Climate and Energy Package 2020 was resolved in 2007 and enacted in legislation in 2009 (Council of the European Union, 2007).

The Ecodesign Directive has been implemented by passing the *Energy-Related Products Law* (EVPG). It provides the framework for improving energy efficiency of specific product groups. The implementation of the Energy Efficiency Directive led to the Energy Service Law and to the definition of the National Action Plan on Energy Efficiency. The *Energy Service Law* (EDL-G) requires large companies to perform energy audits. Exceptions are possible for enterprises which have an energy or environmental management system. In 2014, the government passed the *National Action Plan on Energy Efficiency* as a strategy to focus on energy efficiency activities including the cornerstones energy-efficient buildings, establishing energy efficiency as business model, and increasing responsibility for energy efficiency (*e.g.*, sensitization and transparency) (Bundesministerium für Wirtschaft und Technologie, BMWi, 2014, p. 20).

Further important national legislations include the Energy and Electricity Tax Law, the Energy Saving Directive, and the Renewable Energy Heat Law. The *Energy and Electricity Tax Law* (StromStG) regulates the payment of taxes on energy consumption for end consumers. Furthermore, it includes regulations on exemptions and reductions from this tax. For example, industrial enterprises who maintain an energy or environmental management system may apply for a tax reduction. A prerequisite for this reduction is that the entire manufacturing industry achieves a defined goal on increasing energy efficiency every year.

The *Energy Saving Directive* (EnEV) expresses requirements for residential and non-residential buildings. It regulates the heat transmission coefficients and primary energy demand for buildings including building services (*e.g.*, ventilation). The fulfillment of these standards is required in order to receive a building permit for new or refurbished buildings. The *Renewable Energy Heat Law* (EEWärmeG) requires buildings to use a defined share of renewable energies for heat generation (*e.g.*, solar thermal energy for room heating of residential buildings).

2.2.3 Energy Costs in Industry

The consumption of energy comprises an important share of the total production costs for industrial enterprises. Globally, energy costs account for 12.3 % of the total costs, while this share greatly varies between different industrial sectors (United Nations Industrial Development Organization, 2011, p. 69). For example, the highest share can be found in refined petroleum and nuclear fuel industry with 61.6 %, whereas the lowest share is 0.7 % in office and computing machinery industry (United Nations Industrial Development Organization, 2011, p. 69). Furthermore, the share tends to be higher in developing countries. This may be caused by a lower adaptation of efficient technologies in these countries. Figure 4 shows the share of energy costs in different industrial sectors in Germany.

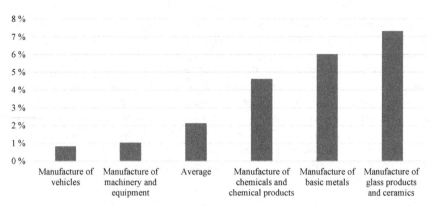

Figure 4: Energy cost's share of total production costs in German industry, divided by sector (data from Statistisches Bundesamt, 2015, pp. 279 ff.)

Although the share of energy costs is low compared to other cost factors (*e.g.*, personnel), it still represents an important lever for increasing profitability of a company. This is especially important when energy prices are presumed to increase in the future. The most commonly used energy carriers in industry are natural gas and electricity (Bundesministerium für Wirtschaft und Technologie, BMWi, 2015). Between 2008 and 2012, energy prices in the European Union annually increased by an average of 1 % for natural gas and 3.5 % for electricity (European Commission, 2014b, p. 5).

The energy price is composed of three elements: The first part reflects the costs of an energy supply company, *i.e.*, for generating energy and delivering it to the grid (energy production). Further costs occur for transmitting energy in an energy network (energy distribution). Finally, taxes and levies are applied according to the governmental policy of a country. For the first element on energy production, prices on wholesale and retail level need to be distinguished: The wholesale price mainly depends on the market structure and may vary several times during the day. The retail costs cover expenses that are required to sell energy to final consumers. While retail energy prices increased as indicated above, the wholesale prices declined by between 35 % and 45 % on the major European wholesale electricity benchmarks during the same period of time. (European Commission, 2014b, p. 6)

The development of electricity retail prices for industry in Germany, separated into production and distribution on the one hand, and taxes and levies on the other hand, is presented in Figure 5. In 2014, energy production costs accounted for approximately 4 to 4.5 Cent and 2 to 2.5 Cent for distribution, while taxes and levies summed up to 8.4 Cent (Bundesverband der Energie- und Wasserwirtschaft, BDEW, 2015, p. 17). The figure shows an increase in the electricity price of 27 % in the past five years between 2010 and 2014.

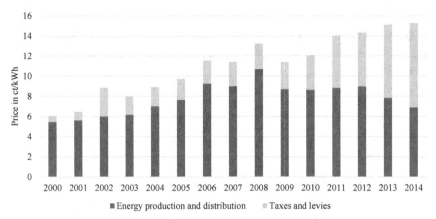

Figure 5: Development of electricity prices in German industry (adapted from Bundesverband der Energie- und Wasserwirtschaft, BDEW, 2015, p. 13)

Taxes and levies contain the following main components: The *electricity tax* is part of the general taxation. The *concession fee* is paid by electricity generating companies to governmental institutions in order to use public area for power lines. The *EEG levy* as part of the Renewable Energy Sources Act (EEG) needs to be paid to operators of energy distribution networks. They use these revenues to pay remunerations to end users that are generating renewable energy (*e.g.*, photovoltaics). The remunerations from the law on renewable energies represent a political incentive for an increasing use of renewable energy. A similar mechanism is pursued by the *combined heat and power levy*, which gives an incentive for the expanded usage of equipment that cogenerates heat and power. (Bundesverband der Energie- und Wasserwirtschaft, BDEW, 2015, p. 13)

The development of electricity prices in the future is mainly driven by the structure of electricity generation and further political decisions. It is expected that electricity prices for industry will rise by 50 % until 2025 due to an increase of EEG levy and wholesale prices (Schlesinger, Lindenberger & Lutz, 2014, p. 227). Therefore, the economic effects of energy consumption pose an important challenge to maintain the competitiveness of industrial enterprises.

2.2.4 Structure of Energy Consumption in Industry

Besides the importance of saving energy, the possibilities to realize these savings need to be considered. Industry causes 28 % of the end energy consumption (Bundesministerium für Wirtschaft und Technologie, BMWi, 2015). This energy is used for several applications (*e.g.*, process heat, mechanical energy).

Figure 6 shows how various applications contribute to the energy consumption in industry. It can be seen that the main usage of energy is for process heat and mechanical energy (*e.g.*, drives), which means that improvement potentials may be especially important in these areas.

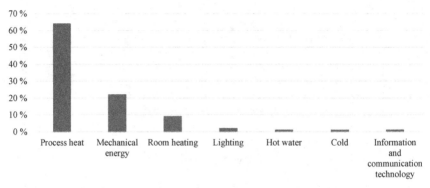

Figure 6: Structure of energy consumption in German industry (data from Bundesministerium für Wirtschaft und Technologie, BMWi, 2015)

It should be noticed that the structure of energy consumption within a factory highly depends on the industrial sector. In general, the *continuous manufacturing industry* is characterized by processing materials and substances, *e.g.*, paper industry, whereas the *discrete manufacturing industry* produces single items, *e.g.*, machinery industry (Chryssolouris, 2006, p. 55). Enterprises in process industries usually have a high energy consumption for generating process heat (*e.g.*, for melting material). The energy consumption of discrete parts manufacturing is as diverse as the products (*e.g.*, automotive, electronic products).

For example, Figure 7 shows the energy consumption structure of the car body shop in an automotive plant. The main process contains manufacturing the car bodies, which requires welding and other production equipment.

Due to the heat generation of welding, the equipment needs to be cooled. Moreover, proper work conditions are maintained, which includes lighting and heating systems. In this example, the heating of the building is realized in combination with the air ventilation system and requires natural gas. Parts of the production equipment need compressed air. Hence, the compressed air is generated with compressors in a centralized supply room and afterwards distributed into the car body shop. The necessary end energy carriers are electricity, water, and natural gas, which need to be purchased.

This example demonstrates the necessity of a holistic consideration in order to increase energy efficiency. For example, a singular improvement measure at the welding system that reduces the electricity consumption in turn effects the heat generation and, hence, the cooling energy demand. Furthermore, the example shows that interdisciplinary action areas need to be considered as part of an energy efficiency project (*e.g.*, manufacturing technology, media supply).

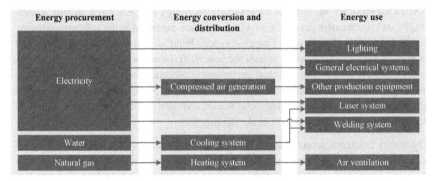

Figure 7: Energy interrelationships in an automotive car body shop (adapted from Engelmann, 2013)

2.2.5 Energy Saving Potentials

The prerequisite to increase energy efficiency are improvement potentials, which can be distinguished into different types: The *theoretical potential* describes possible improvements in contrast to theoretical aspects of physics. It compares, for example, the actual energy consumption of fusion welding with the energy that is theoretically needed to melt the material. The *technical potential* considers technologies that are commercially available. The *economic potential* additionally regards the economic usage of technologies, *i.e.*, it is limited to the cost-effective implementation of a measure. Finally, the *market potential* considers implementation barriers and other market imperfections. The basis of quantifying the market potential is to define a period of time and to estimate a probable scenario including assumptions on which energy efficiency measures are realized. (Schmid et al., 2003, pp. 6 f.)

Several scientific studies analyze the energy efficiency potentials in manufacturing industry. Within these studies, a differentiation is usually made between sector-specific potentials and potentials across several technologies: Whereas sector-specific potentials focus on a branch or specific technology (*e.g.*, paper industry), a high potential can be found in cross-sectional technologies that are applied in a variety of industrial sectors (*e.g.*, drive systems).

In 2008, the Fraunhofer Group for Production conducted a comprehensive study on the potential to increase resource efficiency (Neugebauer et al., 2008). They distinguish improvement potentials according to their realization time into short-term (less than two years), mid-term (two to five years), and long-term (five to ten years) with an increasing intensity of possible changes of a process chain. One result of the study is to quantify the technical potential for manufacturing industry with up to 30 % savings in energy consumption (Neugebauer et al., 2008, p. 344).

BAUERNHANSL ET AL. present a meta study on energy efficiency and include several studies on improvement potentials, of which a few are explained in the following (Bauernhansl et al., 2014). The German Energy Agency (dena) identifies the economic potential of manufacturing industry as 11 % until the year 2020 (Deutsche Energie-Agentur GmbH, dena, 2012, pp. 87 ff.). The technical improvement potential is estimated around 20 % in the long-term (Seefeldt, Berewinkel & Lubetzki, 2009, p. 3). Savings in electrical energy are predominant in cross-sectional technologies and may reach between 80 % and 90 % depending on the scenario (Pehnt et al., 2011, p. 56).

The energy efficiency potential varies between industrial sectors. It is believed that the absolute saving potential is especially high in continuous manufacturing industries, whereas the relative saving potential is higher in discrete manufacturing industries (Bauernhansl et al., 2014, p. 56). This is due to the fact that process industry is rather energy-intensive, which means that energy costs have been focused on earlier. SCHRÖTER ET AL. present a survey to quantify the improvement potential in industrial enterprises (Schröter, Weißfloch & Buschak, 2009, p. 4): Whereas 21 % of the respondents in automotive industry estimate the technical potential above 20 %, this share only accounts for 11 % in the paper industry. A study by the Bavarian Industry Association (vbw) quantifies the economic and technical potential of various discrete manufacturing industries (Table 1).[2]

Table 1: Primary energy saving potentials in different industrial sectors (based on data from Vereinigung der Bayerischen Wirtschaft, vbw, 2012, p. 45)

Sector	Economic potential	Technical potential
Automotive industry	16 %	21 %
Machinery industry	12 %	21 %
Manufacture of basic metals	7 %	11 %
Manufacture of electrical equipment	8 %	22 %

2.3 Barriers for Implementing Energy Efficiency in Industry

Despite the driving concerns and improvement potentials, enterprises face barriers that hinder the implementation of energy efficiency measures. A survey by the Association of German Engineers among 150 industrial enterprises reveals that about 50 % of the respondents have performed an analysis on energy efficiency (Böttger, 2010, p. 46).

---

[2] It should be noted that this study analyzes savings in primary energy, whereas the aforementioned ones consider end energy.

However, in many cases, no measures are deduced from the analysis. Consequently, there are further barriers for implementing energy efficiency even for sensitized enterprises. CAGNO ET AL. categorize barriers against energy efficiency into technology-related, information-related, economic, behavioral, organizational, competence-related, and awareness-related (Cagno et al., 2013, pp. 295 ff.). Technology-related barriers describe the unavailability of energy-efficient technologies, *e.g.*, low diffusion of technologies (Cagno et al., 2013, p. 298). A lack of information can be observed when enterprises do not know about energy efficiency measures or additional information, such as costs and benefits of a measure (Cagno et al., 2013, p. 298). An important obstacle is the missing transparency on the energy consumption within a company (Bauernhansl et al., 2014, p. 103). Economic barriers mainly describe the low availability of capital to realize energy efficiency measures (Cagno et al., 2013, p. 297). Another aspect are internal specifications for short pay-back times (Brüggemann, 2005, p. 35). A behavioral barrier depends on the decision-making actions of an enterprise (Cagno et al., 2013, p. 297); for example, when other objectives are interpreted as being more important.

Organizational criteria contain all aspects of the structural and procedural organization, such as lack of human resources, complex decision chains, or no responsible persons for energy efficiency (Cagno et al., 2013, pp. 297 f.). Barriers related to competences comprise lack of specialized know-how (Cagno et al., 2013, p. 298). Finally, aspects on the awareness mean an ignorance towards the topic energy efficiency (Cagno et al., 2013, p. 298).

A survey among 726 small and medium-sized enterprises (SMEs) of the sectors industry, commerce, and construction analyzes existing barriers (Thamling, Seefeld & Glöckner, 2010). According to the results, the main barriers for implementing energy efficiency are lack of capital and too long pay-back times as well as a scarcity of personnel. Figure 8 shows the importance of various barriers and clusters them into economic, organizational, information-related, and behavioral aspects.

Half of the barriers focuses on economic obstacles, which demonstrates the high importance of these issues. The second-most important aspect are information-related reasons, which include the lack of know-how in general and the lack of special knowledge on energy-saving equipment and technologies. The survey results demonstrate the necessity to provide this kind of know-how to industrial enterprises. Additionally, organizational barriers, especially the lack of time to realize energy efficiency, are relevant.

In a more recent survey, BEY ET AL. identify the clusters of information lack and resource allocation (especially of human resources) as the most important barriers for implementing environmental initiatives (Bey, Hauschild & McAloone, 2013, p. 45).

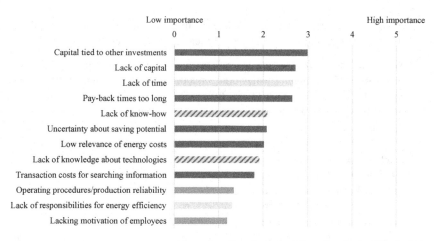

Figure 8: Ranking of barriers towards implementing energy efficiency (Thamling, Seefeld & Glöckner, 2010, p. 27)

The variety of mentioned barriers explains the level of adoption of energy efficiency strategies. Depending on the background of the barriers, different solution approaches are relevant. For example, economic barriers may be addressed by the companies themselves (*e.g.*, availability of capital) or by governmental institutions (*e.g.*, subsidies for energy-efficient technologies).

The development of methods and tools in research especially addresses information-related and organizational barriers. Deduced from the barriers, there is the need to develop methods that provide information on energy efficiency measures (informational aspect) and that can be applied with a manageable effort for finding the relevant information (organizational aspect).

2.4 Interim Conclusion on Energy Usage in Industry

Energy efficiency is an important objective for industrial enterprises. Energy costs represent a significant share of the total production costs, whereof its amount depends on the specific industrial sector. Due to the development of energy prices, especially in terms of increasing taxes and levies, it is assumed that energy costs will increase in the future. Furthermore, energy consumption results in harmful effects on the global ecological system (*e.g.*, carbon dioxide emissions). Against this background, an increasing number of legal requirements obliges industrial enterprises to reduce their energy consumption.

Scientific studies reveal notable potentials to save energy consumption, both within specific industrial sectors and across several sectors. Despite its importance, the topic energy efficiency has not been implemented in depth so far. This is mainly due to economic, organizational, and information-related obstacles regarding the increase of energy efficiency. The empirical results about these barriers point out the need to develop methods and tools that provide information on energy efficiency measures. An important requirement is the handling of these methods and tools without expert knowledge and within a manageable time frame in order to overcome organizational barriers. Since a high number of enterprises estimates that an analysis does not necessarily lead to an improvement, the identification of suitable measures plays an important role.

3 State of the Art

The third chapter presents state of the art that is relevant for the development of a method to identify energy efficiency measures for factory systems. At first, theoretical foundations are explained, which include modeling approaches, systems engineering, and knowledge management. Then, tasks and methods of factory planning are described. Afterwards, approaches to increase energy efficiency in factories are discussed based on scientific and technical publications. Finally, an assessment of the state of the art is conducted in order to concisely define the further research need.

3.1 Models and Modeling

Modeling is a central scientific approach for gaining insights of real objects. This section explains relevant terms and describes various model types. The process to create a model is explained and further illustrated by guiding modeling principles.

3.1.1 Terms and Definitions

A model is defined as a one-to-one mapping of reality that is created for a specific goal (Bamberg, Coenenberg & Krapp, 2012, p. 13). Modeling includes techniques and principles to create a model and is essentially guided by the specific model purpose (Thalheim & Nissen, 2015, p. 35). The main goal of modeling is to describe and analyze an object while reducing the object's complexity (Haußer & Luchko, 2011, p. 3). This reduction in complexity is achieved through abstraction and simplification.

According to STACHOWIAK, a model is described by the characteristics mapping, reduction, and pragmatism: A model is a *mapping* function, *i.e.*, it represents an extract of reality. The *reduction criterion* describes that not every characteristic of the reality is represented in the model, *i.e.*, a model focuses on aspects that are relevant for the desired application. According to the *pragmatic criterion*, a model is conceived and focused upon a specific application case. Therefore, it emphasizes certain aspects depending on the application. This implies that the model is not necessarily universally valid. However, the restrictions of a model need to be stated. (Stachowiak, 1973, pp. 131 ff.)

A model is characterized by a modeling subject, the modeled object that is represented by a corresponding original object, and recipients, for which the model fulfills a specific purpose (Steinmüller, 1993, p. 178). Models are widely used across different scientific disciplines. In general, models are used in science for gaining knowledge, *i.e.*, to describe, explain, and demonstrate an object including its elements and relations (Holzmüller & Bandow, 2010, p. VII). The application in various disciplines induces discipline-specific model types, modeling methods, and tools (Haußer & Luchko, 2011, p. 5).

3.1.2 Model Types

As diverse as the model intentions is the variety of model types, which may be classified as depicted in Table 2. The *measurement level* distinguishes between qualitative models and quantitative models. Qualitative models represent the elements of a system, effects, and purposes. Hence, a qualitative model is most commonly used to detect relationships and developments in a qualitative way, whereas quantitative models are used to calculate values of variables (Hoffmann & Witterstein, 2014, p. 1). An example for a qualitative model is the analysis of information flows and systems in a factory. A quantitative model is, for example, the quantified description of production rate, cycle time, and scrap rate of manufacturing equipment.

Table 2: Classification of model types (Bandte, 2007, p. 205; Cleff, 2015, pp. 10 ff.; Hoffmann & Witterstein, 2014, pp. 1 f.; Schmigalla, 1995, p. 244 f.)

Criterion	Model types
Measurement level	Qualitative models, quantitative models
Level of abstraction	Physical models, visual models, analogous models, formal models
Purpose	Description models, explanatory models, forecast models, decision models, optimization models, simulation models
Spatial dimension	Related to a specific place, not related to a specific place
Temporal dimension	Static models, dynamic models

Models may come in a variety of appearances, which is captured by the *level of abstraction*. Physical models are made out of a solid body, *e.g.*, a true-to-scale plastics model of a machine. Visual models represent objects by their cartesian coordinates, *i.e.*, a two- or three-dimensional graphical representation (*e.g.*, production layout). Analogous models substitute the characteristics of the original object through similar characteristics (Schmigalla, 1995, p. 250). As an example, a layout may be transformed into a structural figure that represents the schematics of machinery without proper scaling. Formal models reveal the highest level of abstraction since they describe the characteristics of an object through mathematical equations or logic symbols (Schmigalla, 1995, p. 251).

Modeling may follow various *purposes*, which is related to the selection of a suitable model type. Description models are mainly used for documentation purposes (Schmigalla, 1995, p. 245), such as reports on quality defects in a production system. The content of a description model is a mere observation, whereas explanatory models contain causes and reasons for a system behavior (Bandte, 2007, p. 205). Forecast models use the explanation on interdependencies to deduce assumptions on the future behavior of a system (Cleff, 2015, p. 12). Decision models are used to prepare the selection of a preferred alternative, such as the calculation of expected savings for energy efficiency investments. Optimization models are usually mathematical representations of optimization problems

(Zimmermann, 2008, pp. 2 f.). This means that they describe a decision situation, in which target objectives are to be minimized or maximized (*e.g.*, the formal description for reducing throughput time depending on machine capacities). The target-oriented manipulation of variables is realized in simulation models. They are primarily used to check the function and operation mode of a developed planning solution (Schmigalla, 1995, p. 254).

The *spatial dimension* of models is used to assign objects to coordinates on a planar surface or in a three-dimensional area, for example to analyze the dimensions of machinery and path widths in a production layout. The *temporal dimension* distinguishes between static models, *i.e.*, which are limited to a specific point of time, and dynamic models including a development over time (*e.g.*, equipment wear during its life cycle). Furthermore, dynamic models may be created by using discrete time or continuous time (Ortlieb et al., 2013, p. 12).

Modeling is interlinked with system theory since modeled objects are often considered as systems; vice versa, systems are usually represented as models (Hopf, 2016, p. 36). Systems and their characteristics are explained in detail in Section 3.2.1. Regarding the modeling of systems, models may be classified with regard to the applied level of abstraction (van den Bosch & van der Klauw, 1994, pp. 43 f.): A "black box"-consideration focuses on the relations between a system and its surrounding but does not consider the internal structure of a system. That is, a black box describes the input and output of a system without describing how the correlation between both works. On the contrary, the "white box"-perspective considers the internal structure of a system. While black box models are developed based on observations and experiments, white box models rely on proven axioms.

The aspects that are represented by models may contain (Kastens & Büning, 2014, p. 22):

- structure,

- characteristics,

- relationships, and

- behavior.

The structure describes the relevant elements of the considered object, such as parts of a product or components of a factory. Characteristics represent properties of an object, which may be either stable (*e.g.*, type of production system) or varying due to changes in the state of an object (*e.g.*, amount of work in progress). Relations may exist between elements of an object, such as exchange of flows (*e.g.*, material transport between two machines) or logical restrictions (*e.g.*, a machine needs to wait for its predecessor to finish a part). Finally, the behavior of an object refers to the transition between states (*e.g.*, processing a new order requires set up).

The elements of a general system may represent different entities, such as organizational units or technical equipment (Kreimeyer & Lindemann, 2011, p. 37). Each kind of an entity represents a system perspective, which is called a *domain*. For example, the Architecture of Integrated Information Systems (ARIS) defines a framework for business modeling including the perspectives organizational structure, data, functions, and processes (Seidlmeier, 2015, p. 57). When representing an organizational structure, the modeled entities include organizational units, persons, groups of persons, positions, and job descriptions.

In this thesis, static and qualitative models are used to represent factory planning tasks and energy efficiency knowledge. Description models are used to represent the factory system, whereas explanatory aspects are represented in the influential parameters on a system's energy consumption. Models are generated in a graphical or formal manner, whereas the latter ones are especially used with regard to the algorithm that assigns energy efficiency measures to factory planning tasks.

3.1.3 Modeling Process and Modeling Principles

The modeling process is depicted in Figure 9. The first step is to create a model from the real object or real problem by developing an abstract representation of reality. As explained above, an appropriate model type and modeling restrictions need to be determined in this step. Afterwards, the model is analyzed, calculated, or modified in order to gain the desired insights. In the third step, the findings are interpreted with regard to the real object. Finally, the findings of the model analysis are transferred to the real object. At this point, the effects of model simplifications need to be examined.

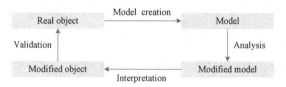

Figure 9: Modeling process (adapted from Ortlieb et al., 2013, p. 5)

As introduced before, the purpose of a model may vary from a mere description to complex forecasting methods. In order to achieve the model's goal, the modeling procedure should be executed in an appropriate quality. Hence, modeling principles and guidelines need to be considered.

BECKER ET AL. define the following general principles of proper modeling, which may be used to assess the quality of model construction (Becker, Rosemann & Schütte, 1995, pp. 437 ff.):

- correctness,

- relevance,

- efficiency,

- clarity,

- comparability, and

- systematic structure.

A model should be *correct* with regard to syntax and semantics. Syntactical correctness means that the model complies with the formal requirements of the applied modeling language, such as the correct usage of symbols (Becker, Probandt & Vering, 2012, p. 32). The model is semantically correct if it behaves as the original system does (Becker, Rosemann & Schütte, 1995, p. 438). This means that there is proper quality of the content in the model and the usage of terms is agreed upon (Becker, Probandt & Vering, 2012, p. 33).

The elements and relations in a model should be *relevant* towards the modeling goal. This means that the modeling goals need to be specified explicitly (Becker, Probandt & Vering, 2012, p. 33). The relevance refers both to the selected extract of reality and to the level of abstraction (Becker, Rosemann & Schütte, 1995, p. 438). Decisions on relevant model components and degrees of detail influence the selection of suitable modeling techniques (Becker, Probandt & Vering, 2012, p. 33).

The effort for the modeling process should be limited in order to make the model *efficient*. Therefore, the intensity and costs for modeling actions are restricted depending on the expected benefit of a model (Becker, Rosemann & Schütte, 1995, p. 438). This refers to the level of detail, modeling techniques and tools as well as to the usage of existing reference models (Becker, Probandt & Vering, 2012, p. 34).

A model should be *clear*, which is especially important for graphical models. This includes a clear structure, readability, and transparency (Becker, Rosemann & Schütte, 1995, p. 438). This criterion is subjective and needs to be adjusted to the individual recipient. Aspects that should be considered include the visual arrangement, filtering possibilities, and hierarchy of passing information (Becker, Probandt & Vering, 2012, p. 35).

The *comparability* of a model emphasizes the necessity that models which are created by different methods should be consistent (Becker, Rosemann & Schütte, 1995, p. 439).

Furthermore, equal aspects of the real object should be identically represented in the model (Becker, Probandt & Vering, 2012, p. 36).

A model can be created with various perspectives on the object by a *systematic* procedure. However, these views need to be integrated in a consistent manner. This requires a common overall structure integrating the various perspectives. Furthermore, aspects that emphasize one of the perspectives should be considered in context to the other viewpoints. (Becker, Rosemann & Schütte, 1995, p. 439)

An example to regard various perspectives on a modeled object is the analysis of business processes in an enterprise. These may be considered from four different perspectives (Schönsleben, 2001, pp. 147 ff.): The *process view* regards the sequence of functions and procedures, while the *function view* clusters tasks according to the accomplished function. The *object view* emphasizes the characteristics of objects, such as organizational units or roles in the enterprise. Finally, the *task view* summarizes functions and processes into intrinsically related tasks, which may be assigned to an organizational unit. Accordingly, modeling procedures may be differentiated into process-oriented, function-oriented, object-oriented, and task-oriented approaches.

The application of the modeling process to a practical problem is a complex task, which is supported by general guiding principles, such as (Ortlieb et al., 2013, pp. 6 ff.):

- precise determination of the real object,

- description of applicable regularities,

- definition of relevant information,

- transfer of existing approaches, and

- identification of appropriate modeling parameters.

Before modeling, the real object or problem needs to be *precisely determined*. This includes defining modeling goals and the relevant system boundaries, *i.e.*, the decision, which components are substantial for the problem and which might be omitted. Afterwards, *applicable regularities* that are valid for the relevant object are identified and described. This may contain theoretical fundamentals, such as physical laws or known interrelationships. *Relevant information* needs to be defined, *i.e.*, it needs to be clarified which information is relevant for the modeling process. This principle helps to reduce the problem to its significant core.

An important consideration for modeling is analogy, *i.e.*, the *transfer of existing approaches*. For example, partial modeling problems may be solved already or studied in other scientific disciplines. Finally, when formalizing a model, the *appropriate modeling parameters* need to be selected. This means stating the relevant parameters depending on the modeling goal and describing the way they are acquired for a practical problem.

3.2 System Theory and Systems Engineering

System theory is a general mindset to describe and understand the structure and behavior of objects. In the following section, characteristics of systems are explained. Afterwards, methods and instruments of systems engineering are presented.

3.2.1 Terms and Definitions

Systems science deals with describing the structure, functionality, and behavior of complex systems (Hitchins, 2007, p. 31). System theory is focused on the behavior of general systems and, hence, it provides approaches that are not restricted to a specific scientific discipline (Ulrich & Probst, 2001, p. 19). The systems perspective is a widely applied methodology and provides several advantages, such as methods and tools for analysis and optimization as well as the methodical description of boundaries, influential factors, and relations (Aggteleky, 1990, p. 226).

A system is defined by the following characteristics (Hitchins, 2007, p. 27):

- holism, *i.e.*, a system is different from the sum of its parts,

- organicism, *i.e.*, the interacting parts within a system behave as a unified whole,

- synthesis, *i.e.*, parts can be formed to something whole,

- variety, *i.e.*, parts of a subsystem complement each other and differ with regard to processes and interactions, and

- emergence, *i.e.*, emergent behavior occurs by interactions between the parts.

Thus, a system can basically be understood as an entity of elements that have specific characteristics and relations among each other (Aggteleky, 1990, p. 225). A system exists for a specific purpose, which means that the elements and their relations are arranged and equipped with various functions in order to achieve a desired system behavior (BKCASE Editorial Board, 2014, pp. 101 f.; Kreimeyer & Lindemann, 2011, p. 36). An important possibility is the hierarchical structure of systems: Systems may be divided into subsystems (Haberfellner et al., 2012, p. 36). By this, systems can be analyzed following a "top-down"approach in order to increase specificity; vice versa, a "bottom-up"approach allows to unite systems for a general consideration.

A system is embedded into an environment, with which it can interact through inputs and outputs that cross the system boundary (Kreimeyer & Lindemann, 2011, p. 36). The delimitation of a system is defined, such that the relation within a system is stronger than between the system and its environment (Haberfellner et al., 2012, p. 35). Depending on the interaction with the surrounding, open and closed systems are distinguished.

Systems sciences are composed of systems thinking and systems engineering (BKCASE Editorial Board, 2014, p. 158). Systems thinking describes basic concepts and principles and, thus, contains the paradigm of how real world phenomena are perceived (BKCASE Editorial Board, 2014, p. 62). Methods and tools of systems engineering are described in the following sections.

3.2.2 Systems Engineering Methods and Tools

Systems engineering is a general approach to apply systems thinking to the life cycle of an engineered system (National Aeronautics and Space Administration, NASA, 2007, p. 3). It describes a methodology to solve problems, whereof problems are defined as a difference between the actual situation and a target situation (Haberfellner et al., 2012, p. 27). The concept is composed of several elements (Figure 10): The systems engineering philosophy provides the mental framework; it contains the principles of systems thinking and the general procedure model of systems engineering. The second component is the problem-solving paradigm that describes a general approach to solve problems. It is accompanied by methods of project management since the solving of systems engineering problems is usually organized in form of projects.

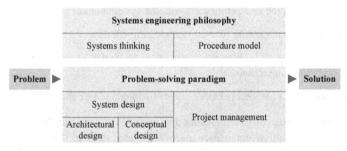

Figure 10: Basic concept of systems engineering (Haberfellner et al., 2012, p. 28)

The procedure of the problem-solving paradigm is depicted in Figure 11 and explained in the following paragraphs (Haberfellner et al., 2012, pp. 74 ff.; Hitchins, 2007, pp. 173 ff.): The first step is to *define the problem space*, *i.e.*, the analysis of the situation in order to identify and understand the problem to be solved. Several perspectives on the situation might be helpful: Based on systems thinking, the situation can be modeled as a system with its elements and the boundary to the system environment. Furthermore, the symptoms of the current situation may serve as basis to identify causes and, with that, deduce suitable solutions. The result of the first step is qualitative and quantitative information on the problem.

Afterwards, two independent activities are carried out: *Identifying ideal solution criteria* means to formulate requirements for a problem solution. These objectives are used for the assessment later on. In addition, *conceiving solution options* is the creative task to generate variants to solve the problem. These concepts should present a level of detail high enough to compare them against each other. This is part of the next step, the *trade-off to find an optimum solution*. It includes checking each variant regarding its suitability (*e.g.*, pre-selection). Furthermore, the variants are assessed and compared based on the solution criteria. (Haberfellner et al., 2012, pp. 74 ff.; Hitchins, 2007, pp. 173 ff.)

Selecting the preferred option has the purpose to define the "ideal" solution variant and is based on the results of the prior assessment. Finally, *strategies and plans to implement* the selected option are formulated. This means to define subsequent measures towards the realization of a variant. (Haberfellner et al., 2012, pp. 74 ff.; Hitchins, 2007, pp. 173 ff.)

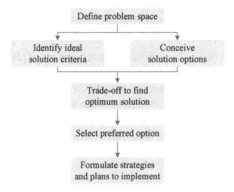

Figure 11: The systems engineering problem-solving paradigm (Hitchins, 2007, p. 173)

3.2.3 Decision-Making Methods

Decision theory is a theoretical approach to support making an "optimal" decision, whereof the definition of objectives for optimality is part of the decision problem. This is an important part of the problem-solving paradigm (see Section 3.2.2).

A decision is defined by the following elements (Nedjah & Macedo Mourelle, 2005, p. 31): There is a finite or infinite set of alternatives and a set of consequences for each alternative. The goal of making a decision is to select a preferred alternative through considering one or more objectives. A typical decision problem occurs when selecting a suitable variant as part of the factory planning process.

Decision-making methods may be distinguished according to the type of problem which they address: As such, methods differ between single and multiple criteria decisions on the one hand, and decisions under certainty and uncertainty on the other hand. In the case of multiple criteria, there is the need to determine preferences between the criteria. If a decision is made under uncertainty, *i.e.*, the consequences of a decision are not completely known, a rule for evaluating the trade-offs between risks and chances is required. (Götze, 2014, p. 45)

Furthermore, monetary and non-monetary assessment methods are distinguished with regard to the optimization objective: Monetary assessment methods evaluate a planning alternative by means of a financial objective, *i.e.*, they quantify cost and revenues of an alternative. Since factory planning projects usually involve long-term investments, methods of investment appraisal are most commonly used. In contrast to cost accounting methods, these consider the effect of payments over several years, *i.e.*, over the life cycle of an investment. Non-monetary assessment methods allow considering qualitative and quantitative objectives, which are not limited to financial factors (*e.g.*, flexibility). Usually, conflicting objectives are part of a decision problem (*e.g.*, trade-off between quality and productivity). Hence, non-monetary assessment methods describe how to solve multiple criteria decision problems. (Götze et al., 2013, pp. 251 f.)

In factory planning projects, the non-monetary assessment of planning alternatives is usually performed by value benefit analysis or point rating (Grundig, 2015, pp. 201 ff.). The principle of the *value benefit analysis* is to define and weigh assessment objectives, which are evaluated in terms of the degree, to which an alternative fulfills the objective. The value benefit of an alternative is then calculated by multiplying the weights with the fulfillment degree and summing up these values for each objective. The main steps of value benefit analysis are (Grundig, 2015, p. 203):

- defining assessment objectives in a hierarchical scheme,

- defining weighting factors for each objective,

- determining partial utility values for each alternative and objective,

- determining overall value benefit for each alternative, and

- deciding for preferred alternative.

The *point rating* method is based on a simplified assessment and decision approach. Within this method, several important assessment objectives are defined, which means that a pre-selection of objectives is necessary. Then, a minimum fulfillment degree is established for each objective. Afterwards, each alternative is assessed whether it does fulfill, partially fulfill, or not fulfill an objective. By means of this assessment, each alternative is given a point rating, which is finally used to rank the alternatives and to select the preferred alternative. (Grundig, 2015, p. 204)

3.3 Knowledge Management

Knowledge management is a general approach for structuring and providing knowledge, such as the energy efficiency information as part of the methodical approach. After a brief explanation of relevant terms, the tasks of knowledge management are described. The focus lies on methods to represent knowledge, which are classified into graphic-based and matrix-based methods.

3.3.1 Terms and Definitions

Describing tasks and methods of knowledge management at first requires the definition of relevant terms. For this, symbols, data, information, and knowledge are considered (Figure 12).

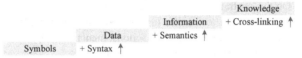

Figure 12: Data, information, and knowledge (adapted from North, 2016, p. 40)

Symbols form the basic collection of signs that can be used to represent superordinate structures (*e.g.*, the alphabet). Data uses these symbols to add syntax, *i.e.*, express facts or statements, such as a name being combined of several letters (Keller & Tergan, 2005, p. 3). Data becomes information when it is given a context and, thus, receives a semantic interpretation, *e.g.*, the information that the name "welding" refers to a manufacturing process (Baets, 2005, p. 90; Krcmar, 2015, pp. 11 f.).

Data and information are classified depending on their measurability. Qualitative information is represented by verbal phrases, whereas quantitative information is based on numerical expressions. Furthermore, quantitative data may be measured on various scale levels (Cleff, 2015, pp. 19 f.): *Nominal* data means to differentiate between several values or groups without any rating (*e.g.*, color of an object). A specific case of the nominal scale is the *dichotomous* scale, in which only the values zero and one may be applied (*e.g.*, maintenance contract is available for a machine or not). A ranking is possible on an *ordinal* scale although the differences between the values do not represent a meaning (*e.g.*, suppliers in A, B, and C categories). The highest scale is the *metric* scale, for which the distance between values has a meaning (*e.g.*, price of machine A is 10,000 Euro higher than of machine B). A metric scale may be measured either *discrete* or *continuously*, depending on whether the characteristic may take any value or is limited to specific values (*e.g.*, the number of employees is limited to the set of integers).

The term knowledge may have various interpretations depending on the discipline that is dealing with this term. With the perspective of business management, knowledge of an organization is defined as the capabilities of the organization to solve problems (Probst, Raub & Romhardt, 2012, p. 22). From a sociological viewpoint, knowledge is a cognitively processed asset that is integrated into existing human knowledge, applied, and modified at the same time (Keller & Tergan, 2005, p. 3). In context of personnel development, learners are supposed to develop knowledge in terms of declarative and procedural knowledge (Niegemann et al., 2008, p. 31). As a consensus on the different disciplines, knowledge is understood as being derived from information by adding understanding to it.

Knowledge can be differentiated into explicit and tacit knowledge. Explicit knowledge may be expressed textually or graphically, hence, it exists in an articulated form, such as in product specifications or manuals (Keller & Tergan, 2005, p. 4). Transferring explicit knowledge is possible with means of information and communication technologies (North, 2016, p. 46). On the contrary, tacit or implicit knowledge is the personal knowledge of an individual and is hard to transfer to other persons (North, 2016, p. 46). The implicit knowledge consists of experiences, actions, mental models, intuitions, beliefs, and emotions of a person (Keller & Tergan, 2005, p. 4).

Knowledge management is defined as the management of the resource knowledge with the goal to enhance the performance of an organization (Schönsleben, 2001, p. 30). The purpose is to get the right knowledge to the right people in the right form at the right time (Schreiber et al., 1999, p. 72). A major challenge in knowledge management is encountered due to the complexity of transferring knowledge (Lehner, 2014, p. 58).

The tasks of knowledge management mainly describe the operative activities of an organization in order to achieve the aforementioned goals. This includes development, improvement, assessment, identification, storage, distribution, use, acquisition, processing, transfer, and presentation (Bahrs, 2007, p. 11; Schönsleben, 2001, pp. 31 ff.). Furthermore, knowledge should be preserved in order to make sure that experiences are maintained within an organization (Probst, Raub & Romhardt, 2012, p. 32). It is necessary to continuously develop and regularly actualize knowledge (North, 2016, p. 54). Besides these tactical and operative tasks, a strategic knowledge management is necessary in an organization in order to define the relevant competencies and knowledge to achieve the organization's strategic objectives.

Three different focus areas of knowledge management can be distinguished (Lehner, 2014, pp. 40 f.): The *human-oriented approach* of knowledge management focuses on the individual as bearer of knowledge. In this context, knowledge management means to support the development of cognitive abilities of an individual and is therefore close to social sciences and human resources management. The *technological approach* emphasizes the technologies that are used to collect, use, and distribute knowledge in an organization. Typical applications occur in computer sciences, such as the realization of expert systems. A cross-linking between the two complimentary perspectives is given by the *integrative approach* of knowledge management. The goal of this holistic approach is to use innovative information technology in order to improve the abilities of employees in an organization.

3.3.2 Knowledge Representation

Knowledge representation is understood as the possibility to visualize knowledge (Keller & Tergan, 2005, p. 2). It supports the externalization of knowledge as well as creating an understanding on knowledge, especially in terms of dealing with its complexity (Olimpo, 2011, p. 93).

Additionally, the development of knowledge representation models allows to connect concepts and to enhance the communication. This supports the collaborative knowledge construction process, *i.e.*, the externalization of knowledge in a group of people (Olimpo, 2011, p. 98). Knowledge representation is the basis to process knowledge (Kasabov, 1996, p. 76).

Depending on the type of the underlying model, there are graphic-based and matrix-based knowledge representation approaches. Graphic-based methods focus either on the formalization of data structures or on the semantics of information. Matrix-based methods represent formal models.

Graphic-Based Knowledge Representation Methods

An intuitive approach to represent semantics is the use of mind maps, which is mainly used to support creative methods, such as brainstorming. In mind mapping, a concept is described based on trees, *i.e.*, a top-down-representation of underlying concepts where there is an unambiguous relationship between one concept and the subordinated concepts. The advantages of using mind maps lie in the ease of use and the support of interdisciplinary communication. (Olimpo, 2011, pp. 105 f.)

More general is the use of concept maps, since they are not necessarily based on a hierarchical structure (Olimpo, 2011, pp. 106 ff.). Concept maps represent knowledge by using two types of nodes: Concepts, usually represented as boxes, may comprise objects, events, and linkages between these concepts (Kramer, 1990, p. 652). Additionally, manifestations of the concepts, *i.e.*, exemplary instances of a concept, can be included (Eppler, 2006, p. 207).

The linkages or relations between concepts in a concept map are not limited to a defined set but need to be specified with the help of verbs, propositions, or phrases (Canas et al., 2005, p. 208). The main advantage of concept maps lies in its emphasis on concepts and their connections (Eppler, 2006, p. 202). This supports the generation of a holistic picture about the central concept that is to be analyzed.

An example of a concept map for the area of compressed air is depicted in Figure 13. Compressors are used to generate the compressed air. Apart from that, the amount of compressed air that needs to be generated, is increased by compressed air losses. These are caused, for example, by leakages and by the efficiency of the compressors. Furthermore, the compressor produces excess heat during operation. This simple example shows the variety of possible concepts (nodes in the concept map), *i.e.*, these might be equipment, energy carriers, or formal parameters. Moreover, the relations between the elements may have varied backgrounds, such as cause-effect-relationships ("leads to") or hierarchical relationships ("is part of"). Hence, concept maps provide an easily comprehensible start for modeling with low formal requirements.

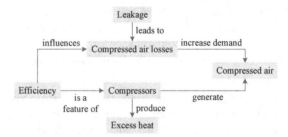

Figure 13: Exemplary concept map (adapted from Schmid, 2004, p. 192)

In contrast to descriptive knowledge, procedural knowledge is represented by graphical business process modeling tools, such as the Structured Analysis and Design Technique (SADT), Petri nets, the Unified Modeling Language (UML), the Architecture of Integrated Information Systems (ARIS), and the Business Process Model and Notation (BPMN) (Gausemeier & Plass, 2014, pp. 248 ff.).

While UML, ARIS, and BPMN provide comprehensive techniques to represent various perspectives on business process modeling, SADT focuses on the core of a process and provides an intuitive access. SADT is a graphical language to model objects and processes (Partsch, 2010, pp. 208 ff.). This method is based on the mathematical graph theory and contains active elements, such as processes and activities, and passive elements, such as data and information (Vogel-Heuser, 2003, p. 103). Passive elements describe the interface to the environment (*e.g.*, information that is generated in a process and transferred to a subsequent process). Each node of the graph is represented by a box and each edge by an arrow.

In the usual form, the so-called activity diagram, activities are nodes and passive elements are edges. In contrast to that, data diagrams contain passive elements as nodes (Partsch, 2010, p. 208). The SADT method supports the hierarchical subdivision of elements, *i.e.*, activities may be partitioned into further sub-activities. An exemplary activity diagram is depicted in Figure 14. The activity receives an *input (I)* and transforms this into an *output (O)* by using a *mechanism (M)*. The mechanism gives additional information on the activity, such as a specific procedure or the involved persons. Finally, the *control (C)* describes data that controls or initiates the activity, *e.g.*, sensor information.

Advantages of the SADT method are especially the support of a top-down approach and its intuitive understanding through the use of natural language. However, checking consistency and completeness is not methodically supported (Partsch, 2010, p. 211).

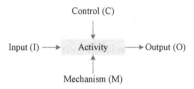

Figure 14: Exemplary SADT activity diagram (Vogel-Heuser, 2003, p. 104)

Matrix-Based Knowledge Representation Methods

The second group of knowledge representation methods is based on matrices, following the purpose to prepare the processing of knowledge. These methods are mainly applied in information technology in order to represent logic rules of expert systems. Moreover, matrix-based knowledge representation approaches are used in production engineering to generally describe connections between different concepts.

The method of design structure matrices (DSM) has been developed in order to manage the design of complex systems (Steward, 1981). In the initial application, the matrix refers to a complex product and the entries represent relations of precedence between variables that define the design of the product. Therefore, the DSM helps to identify interrelationships between single characteristics of product design. With this information, the effect of changes on other parts of the design project can be analyzed (Steward, 1981, p. 71). The concept may be applied to other areas, as well, for example project planning, systems engineering, and strategic management (Browning, 2001, p. 292).

The general notation of a DSM is a quadratic matrix with identical row and column labels. These labels represent the elements in the considered system (*e.g.*, components of a complex product when applied in product development). The entries of the matrix show the dependencies between these elements, *i.e.*, the entries within the row of an element show its inputs from other elements. (Danilovic & Browning, 2007, p. 302)

The content that is represented with a DSM is deterministic, *i.e.*, the elements and the relationships between them have to be known prior to modeling (Kreimeyer & Lindemann, 2011, p. 51). DSMs are differentiated according to the represented type of relationship. Thus, directed and non-directed DSMs on the one hand, and binary and numerical DSMs on the other hand are distinguished (Kreimeyer & Lindemann, 2011, p. 45): In a directed DSM, the direction of a relation between two elements matters, whereas there is no such distinction for non-directed DSMs. The flow of information is an example for a directed DSM (*i.e.*, information output of one process represents the input for the next process). A possible geometrical collision between two components within a product is an example for a non-directed DSM, since this relation affects both components in the same way without a specific direction. The difference between binary and numerical DSMs is that the entries in a binary DSM are limited to values of zero and one.

A relation between two elements in a DSM is comprehensively defined by the following criteria (Tang et al., 2010, pp. 160 f.): The *initiating element* and *affected element* represent the two elements that are in relation with each other. The *type of interaction* specifies the kind of relation that exists between the elements. For example, the type of interaction can be the exchange of information or material between processes or the spatial interaction between two parts in a product. The *level of interaction* refers to the intensity of the relation. This can be expressed on an ordinal or ratio scale (*e.g.*, "low", "high", percentage value). The *milestone* defines the stage in the life cycle of product development, at which the relation needs to be considered. The *criterion* is the objective to evaluate the interaction (*e.g.*, cost, time, performance). (Tang et al., 2010, pp. 160 f.)

One advantage of matrix-based knowledge representation is the fact that the readability does not change with the size of the modeled system. Moreover, a variety of analytical methods may be applied. That means, after identifying the relations between elements in a project, several strategies can be applied in order to improve the performance of the system. For example, avoiding overlaps between different tasks of a project reduces the communication effort between the project participants.

As introduced before, DSMs focus on the modeling of one domain, *i.e.*, one type of elements within a system. Modeling relations between elements of different types (*e.g.*, assignment of persons to tasks) requires considering different domains. The methodical extension of DSMs that can take into account different domains are domain mapping matrices (DMM) (Danilovic & Sandkull, 2005, p. 194). Since the elements originate from different domains, DMMs are in general rectangular in contrast to quadratic DSMs. An example for a DMM is given by:

$$
DMM = \begin{matrix} & P_1\ P_2\ \dots\ P_m \\ \begin{matrix} U_1 \\ U_2 \\ \vdots \\ U_n \end{matrix} & \begin{pmatrix} 1 & 0 & \dots & 2 \\ 2 & 1 & \dots & 3 \\ \vdots & \vdots & & \vdots \\ 0 & 3 & \dots & 0 \end{pmatrix} \end{matrix} , \qquad (3.1)
$$

which represents the relation between the domains P and U (*e.g.*, processes and organizational units). The elements in the matrix characterize the relation, for example the degree to which an organizational unit is involved in a process. In this example, the values of the matrix entries are on an ordinal scale.

A combination of the aforementioned concepts can be done by using multiple-domain matrices (MDM) (Kreimeyer & Lindemann, 2011, pp. 46 ff.): These consist of a variable number of DSMs and DMMs, whereof DSMs are arranged along the diagonal and DMMs outside of the diagonal. With the use of MDMs, modeling of multiple network structures is possible. That means, relations both within a domain and among different domains can be represented. The use of different types of relations is also possible (König, Kreimeyer & Braun, 2008, p. 232).

3.4 Factory Planning and Factory Management

This section presents the state of the art in factory planning and factory management. This includes explaining relevant terms and the understanding of factories as systems. Afterwards, factory planning tasks and approaches are described. The goals of this section are to describe the object area and to explain approaches that may be used to support the method development in this thesis.

3.4.1 Terms and Definitions

A factory can be considered as a "place where value is created by the manufacture of industrial goods based on division of labor while utilizing production factors" (VDI 5200, Part 1, p. 3). These production factors include ground, staff, capital, energy, and information (Felix, 1998, p. 33). A factory is characterized by an organizational structure which is managed both technically and economically (Kettner, Schmidt & Greim, 1984, p. 1). Factories are built and used for a specific time-limited purpose and follow a factory life cycle (Schenk, Wirth & Müller, 2014, p. 150). The main purpose is to manufacture products and commercially offer them to a market (Kettner, Schmidt & Greim, 1984, p. 1).

Other important terms that are used in this context are manufacturing system and production system. While some authors use the term manufacturing system synonymously to a factory or plant (Bellgran & Säfsten, 2010, pp. 43 ff.), the more precise approach is to consider it as the elements that are involved in the physical making of a product (Chryssolouris, 2006, p. 332). Accordingly, the manufacturing is considered as the subsystem of a factory that produces individual components in contrast to the assembly system, in which individual components are joint to produce the final product (Bellgran & Säfsten, 2010, p. 45). The term production system may be understood as the combination of manufacturing and assembly system (Hitomi, 1996, p. 48) or as a framework to set standards for business processes in a manufacturing company, such as the Toyota Production System (Clarke, 2005, pp. 88 ff.).

Factory planning is understood as a "systematic, objective-oriented process for planning a factory [...] and covers all activities from the setting of objectives to the start of production" (VDI 5200, Part 1, p. 3). Factory planning in the narrow understanding has the task to plan a factory including all functions associated with production. In a broader sense, it further comprises the definition of factory and project objectives, the location selection as well as the planning of external logistics. Furthermore, it can also include issues of factory operation, *i.e.*, during the production phase. (VDI 5200, Part 1, pp. 4 ff.)

This wide field of application leads to a high complexity and diversity of planning tasks. To handle this complexity, factory planning is supported by various planning methods, instruments, and tools.

3.4.2 Factory Systems

A common basis to describe factories is the system theory (see Section 3.2.1). The *elements* in a factory system are the basic production factors equipment, material, and personnel (Schenk, Wirth & Müller, 2014, p. 122). The equipment contains the technical systems, such as machinery and other devices. Material comprises the raw material that is used to manufacture a product and auxiliary material that does not enter the product (*e.g.*, lubricants). The personnel fulfills tasks by using the equipment.

Relations describe the connection of elements within the factory system, whereas *structures* link two elements, such as the transport from one machine to another machine, and *processes* link more than two elements, such as the product flow through several machines (Schenk, Wirth & Müller, 2014, p. 122). The relations can be characterized as, for example, information, material, energy, capital, and personnel flows (Schenk, Wirth & Müller, 2010, p. 13).

The environment of a factory is characterized by, among others, natural, infrastructural, economic, and political aspects (Schenk, Wirth & Müller, 2014, p. 123). Figure 15 depicts a simple system representation of a factory. The elements are, for example, machines or logistics equipment. The input to the factory contains material, energy, and information, whereas the output represents products and waste. A subsystem is the composition of elements according to a common characteristic (*e.g.*, group of turning centers).

Technical systems may be modeled with regard to function, structure, or hierarchy (Ropohl, 2009, pp. 75 ff.): The functional view describes input to and output from a system, whereas the structural model considers the internal system structure, *i.e.*, the elements and their relations. The hierarchical perspective addresses the hierarchical relations between system elements. The hierarchical model of a factory system allows gradual consideration of various levels and to combine these with the responsible actors (Müller et al., 2009, p. 41).

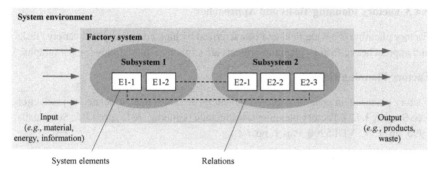

Figure 15: Simplified system representation of a factory (adapted from Schenk, Wirth & Müller, 2014, p. 123)

A manifestation of the functional perspective is to model systems in a factory in a peripheral structure, whereof the layers are determined by the functional importance for the production process. As such, the center is represented by the main processes which are performed by the production equipment. In the first periphery are supporting processes that directly depend on the production program (*e.g.*, storage). The second periphery contains systems that depend on the main process, such as maintenance, whereas the systems in the third periphery are independent from the production process, such as sanitary facilities. (Müller et al., 2009, pp. 43 f.)

Additionally, it is important to consider factories as socio-technical systems, which means that the function of a factory is realized by both humans and technical resources (Schenk, Wirth & Müller, 2014, p. 49). A socio-technical system is understood as the combination of technical and social systems (Krüger, 2007, p. 49): The technical system contains the equipment and resources, the production processes, their necessary conditions, and the physical surrounding conditions. The social system is represented by the members of the organization, which includes their qualifications, individual needs, and group-specific needs. A socio-technical system represents a unit of technical and social systems, which together realize an action (Ropohl, 2009, p. 141). This action consists of a certain goal, required and available information as well as the realization of the action (Ropohl, 2009, p. 138).

When planning a factory as a socio-technical system, it is important to design the interface between factory planning and organization planning. This includes the definition of the operational and organizational structure (Schulze, Reinema & Nyhuis, 2012, p. 213). The relation between the technical and social subsystem is modeled by defining tasks and roles, *i.e.*, assigning responsibilities to each task (Krüger, 2007, p. 50).

3.4.3 Factory Planning Tasks and Approaches

Factory planning is a wide field and characterized by high complexity and variety. Tasks and approaches of factory planning projects are explained in the following paragraphs.

Factory Planning Tasks

Factory planning, in general, comprises planning activities that are related to production (see Section 3.4.1). Factory planning tasks include (Grundig, 2015, pp. 11 f.; Schmigalla, 1995, pp. 13 f.; VDI 5200, Part 1, pp. 6 f.):

- planning of the factory location,

- building planning (including architecture, design, and layout),

- planning of production processes, machinery, and equipment,

- planning of peripheral processes (*e.g.*, media supply, maintenance),

- material flow and logistics planning, and

- staff planning.

The planning tasks can be distinguished into long-term, mid-term, and short-term depending on their timely horizon: Long-term tasks focus on a period of time that is longer than the life cycle of the production equipment (*e.g.*, selection of factory location). Mid-term tasks handle the production equipment (*e.g.*, planning of new machinery). Short-term tasks are realized in shorter time periods (*e.g.*, increase productivity of existing equipment). (Aggteleky, 1987, p. 29)

The timespan of planning tasks is related to the life cycles that need to be considered within a factory. SCHENK ET AL. discuss the interferences between the life cycle of the product, production process, building, and area that need to be managed as part of a life cycle-oriented factory management (Schenk, Wirth & Müller, 2014, pp. 147 ff.).

The sequence of planning activities is described by procedure models for factory planning. These describe the factory planning process in several stages that are executed in a mixture of stepwise and iterative manners. Some authors consider the factory planning process as terminated with the built factory or start of production. Others include the management of an existing factory in order to emphasize the interrelationships within a factory life cycle.

Factory Planning Phases

The factory planning process is composed of seven planning phases (Figure 16). At the beginning, the *setting of objectives* analyzes the goals of the company and the restrictions for the project in order to define the project objectives and the project plan (VDI 5200, Part 1, p. 9). Goals may include technical, economic, temporal, and organizational aspects (Aggteleky, 1990, p. 356). Furthermore, the degrees of freedom for the subsequent planning activities are defined.

Factory planning tasks are characterized by their uniqueness, for example in terms of project budget, system complexity, or planning objectives (Schenk, Wirth & Müller, 2014, p. 288). Therefore, typical tasks and tools of project management need to be applied, such as the definition of work packages (VDI 5200, Part 1, p. 11). Important results of this first phase are the problem description, project schedule, objectives, time, budget, and the project organization (Grundig, 2015, p. 56).

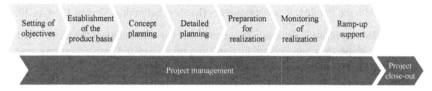

Figure 16: Phase model of the factory planning process (VDI 5200, Part 1, p. 9)

In the second phase, the *establishment of the project basis*, required data and information are collected or generated. This includes information on the product, production (*i.e.*, processes and resources), and building (VDI 5200, Part 1, p. 11). The purpose of this detailed situation analysis is to determine a basis for the later improvement (Grundig, 2015, p. 57). Methods to acquire data may be either direct, *i.e.*, from the process, or indirect, *i.e.*, from existing documentation (Grundig, 2015, p. 60). Examples for data sources are measurements, surveys, observations, building plans, or production statistics (Grundig, 2015, pp. 60 f.).

Within the *concept planning*, a solution concept to meet the objectives is developed. Usually, several variants are conceived and assessed. Assessment criteria may contain the material flow, extension options, work conditions, flexibility, and investment and operating expenses (Kettner, Schmidt & Greim, 1984, p. 26). It should be noted that – if based on an existing system – the ideal concept does not only include improvements; instead, it is based on the functional requirements of the production process (Kettner, Schmidt & Greim, 1984, p. 20). The result of the concept planning is a qualitatively and quantitatively assessed solution variant for the factory system (VDI 5200, Part 1, p. 12).

In the fourth phase, the *detailed planning*, the selected alternative is planned out. The purpose of this phase is to work out details for the rough concept in order to get it ready for realization (Grundig, 2015, p. 208). It includes creating the fine layout, final building plan, and project cost calculation (VDI 5200, Part 1, pp. 15 ff.).

The *preparation for realization* contains preliminary activities for the organizational, technical, and constructional realization (Grundig, 2015, p. 217). During this phase, the project management for the realization is specified including the nomination of a project manager and the project structure planning (Grundig, 2015, p. 217). Tenders are generated and orders are placed for services that are delivered externally (VDI 5200, Part 1, p. 17). Eventually, suppliers are identified and the final, ready-to-build plans are developed.

Afterwards, the *monitoring of realization* is conducted while the construction is executed. This means that the implementation processes are coordinated, monitored, and documented (VDI 5200, Part 1, p. 19). In parallel, first tests are executed, which may be combined with the training of new employees (Kettner, Schmidt & Greim, 1984, p. 30).

The final planning phase is the *ramp-up support*. The commissioning of the factory starts with a pilot phase (test and pre-production), run-up phase, and ends when processes are realized stably under series conditions (Grundig, 2015, pp. 220 f.). The phase also contains the evaluation of the factory according to the initial objectives (VDI 5200, Part 1, p. 21). Eventually, the planning project is concluded and documented.

Factory planning procedures are complex and require the cooperation of several disciplines. These tasks are usually conducted by means of interdisciplinary projects between factory planners and specialist planners. The factory planners are responsible for a holistic consideration of the factory system. They break down subtasks and integrate other participants. Moreover, specialist planners work on the detailed design of certain parts of the factory (*e.g.*, heating system). For example, the German fee structure for Architects and Engineers (HOAI) defines the scope of services of architects and engineers in a building project. Therefore, the tasks of specialist planners focus on building, interior spaces, and technical equipment.

The planning phases refer to the entire life cycle of a factory, which is divided into development, setup, start-up, operation, and dismantling (Schenk, Wirth & Müller, 2010, p. 19): The development phase is equivalent to the factory planning until the step of detailed planning. Setup includes planning the factory implementation and configuring the factory systems. During the start-up, the factory systems are ramped up until they reach normal operation, which marks the shift to the life cycle phase operation. Afterwards, the factory is shut down during the dismantling phase. A more detailed consideration of the factory life cycle is discussed in Section 4.5.

Factory Planning Steps

The particular content within the planning phases is defined by the factory planning steps. These are mainly assigned to the phase concept planning (VDI 5200, Part 1, pp. 12 ff.) and may be conducted following a varying level of detail. The basic planning steps are (Grundig, 2015, pp. 45 f.; VDI 5200, Part 1, pp. 12 ff.):

- determination of functions (structure planning),

- dimensioning,

- structuring (ideal planning), and

- design (real planning).

The *determination of functions* conceives the processes and equipment in a qualitative manner based on the analysis of the production program. It includes all planning activities towards the processes in the factory system model in Section 3.4.2, *i.e.*, the system elements and their relations, such as material, information, and energy flows (Schenk, Wirth & Müller, 2014, p. 301). This means that production processes are selected and arranged as a sequential process chain. The findings of this step contain the type of functional units, their qualitative linkings through the material flow, the chronological structure and the required resources (Grundig, 2015, pp. 80 f.).

The *dimensioning* deals with the quantitative determination of processes and equipment, *i.e.*, the amounts of equipment, media requirements, necessary area, and workforce (Grundig, 2015, p. 88). Within this step, the calculations are based on the balance between installed capacity and expected load (Schenk, Wirth & Müller, 2014, p. 308). For example, the capacity of machines to manufacture parts needs to be higher than the required capacity for the production program. Dimensioning the area accounts for equipment, workplaces, transport, temporary storage, maintenance, quality control, and media supply (Grundig, 2015, pp. 103 ff.). The media requirements mainly contain electricity, heating, ventilation, air conditioning, industrial gases, compressed air, water, and waste water (Grundig, 2015, pp. 109 f.).

Structuring determines the spatial and temporal arrangement of the previously identified functional units (Grundig, 2015, p. 111). Since each manufacturing process is character- ized by a spatial and temporal structure, a suitable organization for the production needs to be selected when the processes are combined (Schenk, Wirth & Müller, 2014, p. 320). Within this step, the material flow dominates the optimization in order to reduce logistic costs (Grundig, 2015, p. 113). The result is an ideal layout including the variants for the building envelope (VDI 5200, Part 1, p. 14).

Within the step *design*, the real layout is developed. Layout and building variants are generated and evaluated by considering given restrictions, *e.g.*, due to an existing building structure (Grundig, 2015, p. 168). Common instruments to support this step are computer-aided design tools, which allow an interactive planning (Grundig, 2015, p. 177). Furthermore, logistics systems are selected and assigned to the corresponding functional units within the factory. The result of the design step is a planned variant that is assessed against the initially determined criteria. Selecting the preferred variant is supported by decision-making methods (Section 3.2.3).

Additionally to these four steps, the interpretation of the product and service program may be considered as first step before the determination of functions (Schenk, Wirth & Müller, 2014, p. 294). This step means to analyze the production program and to define reference products which may represent a number of product variants. The production program describes the performance of a production system with regard to assortment, amount, time, and value of products (Grundig, 2015, p. 64). The product and service program is an important input for factory planning since the function, dimension, and structure of a factory system is determined significantly by the product program (Grundig, 2015, pp. 64 f.).

Factory Planning Cases

The reasons to initiate a factory planning project are as diverse as the planning tasks. In a rough classification, internal and external reasons can be identified (VDI 5200, Part 1, p. 4): Internal reasons occur inside a company (*e.g.*, market strategies, new product variants), whereas external reasons arise from outside the company (*e.g.*, legal requirements). The internal causes can be attributed within the factory (*e.g.*, new production technologies) or the company (*e.g.*, new company strategy). Examples for external reasons are new requirements for the production that are defined by law or changes in the market situation.

Depending on the initial situation, factory planning projects can be categorized into several planning cases. A *new planning* project means to plan a new factory. The restrictions that need to be considered are limited to the terrain and available infrastructure (VDI 5200, Part 1, p. 4). This planning case has the highest degree of freedom (Grundig, 2015, p. 18). Planning a new factory is possible for a new product but it is more common for shifting an existing location (Wiendahl, 1996, p. 9-2).

The *reconfiguration* planning belongs to the more frequent projects. It varies from simple rationalization measures (*e.g.*, cost reduction with existing equipment) to a redesign of the entire factory. Within this category, expansion and shrinking represent specific planning cases. An *expansion* means to increase the capacity of an existing factory due to new products or an increasing production volume (Grundig, 2015, p. 19).

Shrinking means to adjust an existing factory to decreased capacity, for example because of reduced sales figures or an outsourcing strategy (Grundig, 2015, p. 19). The shrinking planning belongs to the most difficult planning cases because the peripheral systems (*e.g.*, energy supply) need to be adjusted to the new requirements (Wiendahl, 1996, p. 9-2).

A *demolition* project means to shut down a factory and dismantle or demolish it in order to prepare the site for a different further use (VDI 5200, Part 1, p. 4). The *revitalization* means to make an industrial wasteland site available for re-utilization (VDI 5200, Part 1, p. 4).

Factory Planning Principles

The factory planning process is guided by general principles which are part of the planning method: The *top down* principle means a gradual increase of detail along the planning process (*e.g.*, starting with a general problem and developing it towards a more detailed one). The purpose of this approach is to keep the overview of the entire problem and to specify it, where necessary. The opposite approach is the *bottom up* principle which means to generalize from a specific aspect (*e.g.*, aggregate detailed information). (Schmigalla, 1995, pp. 89 ff.)

Another perspective is the principle *from outside to inside* which means to consider outside requirements at first (*e.g.*, from the sales market). The principle *from central to peripheral* is the opposite and means to conceive from the production processes to the peripheral processes. The principle *from ideal to real* describes the stepwise integration of restrictions and boundaries. The purpose of this approach is to consider an ideal solution and to identify the aspects that hinder its implementation. Finally, the principle *optimize and vary* summarizes the approach of generating several alternatives in order to increase the variety of concepts. (Schmigalla, 1995, pp. 89 ff.)

3.5 Approaches to Increase Energy Efficiency in Factories

This section presents methods and tools that can be applied to increase energy efficiency in factories. At first, norms and standards in the area of energy-efficient production and energy management are described. Afterwards, instruments that include cooperations between industrial enterprises or commercially available offers as well as generalized energy efficiency principles are presented. Emphasis is put on the explanation of systematic procedures to increase energy efficiency. At the end of the section, the existing approaches are assessed with regard to the goals of the thesis.

3.5.1 Norms and Standards

This section describes norms and standards that are relevant for increasing energy efficiency in industry (Table 3). The international standard on energy management and energy management systems describes the requirements to implement an energy management system within an organization (DIN EN ISO 50001). It resembles the structure of environmental management systems (DIN EN ISO 14001) and quality management systems (DIN EN ISO 9001).

Table 3: Overview of norms and standards to identify energy saving potentials in industry (selection)

Number	Title
VDI 3922	Energy Consulting for Industry and Business
DIN V 18599	Energy Efficiency of Buildings – Calculation of the Net, Final, and Primary Energy Demand for Heating, Cooling, Ventilation, Domestic Hot Water, and Lighting
DIN EN 15900	Energy Efficiency Services – Definitions and Requirements
DIN EN 16231	Energy Efficiency Benchmarking Methodology
DIN EN 16247	Energy Audits
DIN EN ISO 50001	Energy Management Systems – Requirements with Guidance for Use

The standard on energy management contains steps to be conducted by an organization to implement a strategic energy management process. The overall goal is defined as energy-related performance and may contain the objectives energy efficiency, energy use, and energy consumption (DIN EN ISO 50001, p. 9). It follows a general cycle for continuous improvement processes that consists of the steps plan, do, check, and act (PDCA cycle). It starts with an analysis of the situation including the acquisition of available data. Afterwards, a plan for improvement is formulated based on the analysis results. The next step is to implement the plan, after which it is checked whether the measures achieved the desired improvements. The final action step may contain corrections on the improvement measures or a methodological standardization in order to secure the continuous realization of the new practices. (Kamiske, 2015, pp. 142 f.)

The main support lies in the instruments to structure the organization and processes for increasing efficiency. Hence, it can be considered as a management instrument on a general level. There are various interrelationships between energy management and factory planning, since the analysis and evaluation of indicators on energy efficiency represent important input information for factory planning processes (Müller et al., 2013a, p. 626). The identification of energy efficiency measures is an important component for a continuous improvement process. However, the standard does not provide support about how this identification may be conducted.

A more operative approach for energy efficiency improvements is described as energy audit in DIN EN 16247. An energy audit is defined as the systematic inspection and analysis of the energy consumption in order to identify improvement potentials. It can be conducted for single machines, buildings, or entire organizations. The steps of an energy audit are kick-off, data acquisition, on-site visit, analysis, reporting, and concluding discussion. The identification and evaluation of measures to increase the energy efficiency is part of the analysis. The norm describes the information that the energy auditor should provide for the organization (*e.g.*, necessary investments to realize a measure). However, it does not describe how to identify measures. (DIN EN 16247, Part 1)

The VDI guideline 3922 describes requirements and approaches for conducting energy consulting services in industry. This guideline describes an approach that is similar to the energy audit but directly focuses on the improvement of energy efficiency. It consists of the steps establishing contact, quotation and contract, ascertaining the current situation, presenting and assessing the current situation, proposals for efficient energy use, development of overall concepts, assessment and selection of measures, presentation and consultation report, and implementation and efficiency review. (VDI 3922)

As a tool to support the identification of measures, the following types of improvement opportunities are described (VDI 3922, pp. 12 f.):

- avoidance of unnecessary energy consumption (*e.g.*, reduce idling),

- reducing the specific energy consumption (*e.g.*, energy-efficient technologies),

- improving the efficiency and utilization ratio (*e.g.*, reduce distribution losses),

- energy recovery (*e.g.*, heat recovery), and

- use of regenerative energy sources (*e.g.*, solar hot water systems).

The assessment of energy efficiency is supported by a framework for conducting energy efficiency benchmarks (DIN EN 16231). A benchmarking is a quantitative comparison between several units in order to identify the best performance (Alter, 2013, p. 155). This also supports the detection of gaps and, thus, deducing improvement measures. The units to be compared may be diverse (*e.g.*, companies, factories, processes, products). The standard states the requirements for an energy efficiency benchmarking and defines its steps, namely planning, data acquisition, evaluation, and reporting (DIN EN 16231, pp. 9 ff.). The result of the benchmarking approach is the interpreted quantitative data on energy consumption. However, the standard does not describe how this data can be used to identify energy efficiency measures.

The DIN EN 15900 describes requirements for energy efficiency services which include the identification, selection, and realization of energy efficiency measures. This standard structures the approach to provide an energy efficiency service and, hence, helps providers and customers to find a common understanding as basis to sign a contract. For example, it includes the need to quantify the initial situation and to verify the energy savings after the implementation of measures. (DIN EN 15900)

The DIN V 18599 describes how to calculate the energy consumption of a building. It is explained in Section 4.4.2 with regard to energy efficiency of buildings. (DIN V 18599, Part 1)

3.5.2 Industrial Cooperations and Commercial Offers

This section explains opportunities to use external information on energy efficiency measures through industrial cooperations and commercial offers. As a first possibility, companies can exchange their know-how on energy efficiency measures among each other, for example during conferences or within special cooperative networks. An example is the initiative "Energy efficiency networks" of the German Federal Ministry of Economics and Technology that funds regional cooperations between 8 to 15 enterprises. For instance, the network in Heilbronn-Franconia achieved an increase in energy efficiency of 6.4 % between 2009 and 2012 (Betrieblicher Umweltschutz in Baden-Württemberg, BUBW, 2014).

Besides the individual exchange, information platforms about examples of realized energy efficiency measures serve as a central starting point. For example, the German Energy Agency provides a database of industrial energy efficiency projects (Deutsche Energie-Agentur GmbH, dena, 2016). A user of this platform may search projects based on the criteria country, technology, and sector. Each project is described by a brief overview of the implemented measures, the energy and cost savings and a detailed explanation that may be supplemented by photographs.

Guidebooks on special topics (*e.g.*, compressed air) or for specific sectors (*e.g.*, energy efficiency in bakeries) that explain improvement measures are provided by governmental and other institutions (*e.g.*, energy agency, environmental protection agency). By including industrial case studies, the guidebooks represent support that is close to practical application. However, finding the information that is relevant for a use case requires high effort and time. Moreover, the challenge is to transfer this general information to a specific case.

Other information sources are commercially available through consultancies or manufacturers of energy-efficient technologies. Consultancies offer the service to analyze the energy consumption of an industrial enterprise. In many cases, however, energy consultants are experienced with general technologies rather than with the industrial sector of the client. This leads to the fact that the suggestions on energy efficiency measures focus on the building and building services, with a specific emphasis on lighting and heating (Frahm et al., 2010, p. 55). However, studies show that the highest savings can be achieved by realizing technical measures in a coordinated and comprehensive manner (Bründl et al., 2012, p. 45).

Research institutions also offer services to increase energy efficiency. These range from providing self-assessment tools to conducting methodical analyses in a factory. One example is the ecofactory quickcheck (Reinema, Mersmann & Nyhuis, 2011): With the help of an internet tool, an industrial enterprise can do a self-assessment for the areas location, building, process, building services, and organization. An automatic evaluation highlights these areas depending on the improvement opportunities in green, yellow, or red (ecofabrik, 2016). The questions of the tool give an advice where to start in order to exploit saving potentials.

Manufacturers of energy-efficient products usually offer decision support, for example by means of calculation tools. For instance, SIEMENS provides a tool to calculate energy saving potentials and amortization of motors, pumps, and ventilators for specific applications (Siemens AG, 2016). The tool is free of charge and allows to compare the energy consumption between two objects, *e.g.*, a high efficiency motor compared to an older standard motor.

The main advantage of external consultancy is the significant experience of consultants, from which customers can benefit. However, there are some disadvantages as well: In a survey about barriers for implementing resource efficiency, companies mention common obstacles when working with consultancies, namely costs, lack of know-how about the industrial sector, and bad experiences in the past (VDI Zentrum Ressourceneffizienz GmbH, 2011, p. 25).

The strategical integration of energy efficiency into an organization is a crucial factor in order to achieve long-term success (Hirzel, Sontag & Rohde, 2011, pp. 11 f.; Schepers et al., 2013, p. 267). This underlines the necessity for enterprises to build up their own energy efficiency knowledge, which can be supported by consultants and other external information sources.

3.5.3 Generalized Energy Efficiency Principles

Consultancies and research institutions develop general principles in order to structure the identification of measures (Müller, Krones & Strauch, 2013, p. 1629). These principles describe fields of action that need to be considered to optimize energy efficiency within a company. Thus, they help to guide the process of identifying specific measures.

Energy efficiency principles can be regarded as checklists to examine which general guidelines are already considered in a factory system and which still need to be considered. Their structure may be either hierarchical or parallel. A widely spread hierarchical methodology is called the 3 R's – reduce, reuse, and recycle (United States Environmental Protection Agency, 2016b). It means that, at first, energy consumption should be reduced. If this is not possible, waste or lost energy should be reused or recycled. An extension of this approach is given by the 6 R methodology that also includes recover, redesign, and remanufacture (Jayal et al., 2010, p. 151). This concept considers the effects over the entire product life cycle, for example by emphasizing the need to redesign a product to improve its ability for recycling.

A hierarchy of principles that is focused on energy efficiency are the so-called "stairs of innovation": This structure distinguishes energy efficiency principles according to the extent of changes on the machinery. The categories substitution, upgrading, reenginering, product design, and technology management are defined. *Substitution* refers to replacing a single hardware component, whereas *upgrading* means a change of organization or equipment software (*e.g.*, machine control). *Reengineering* comprises minor hardware changes, *e.g.*, new dimensioning of a drive, while a *product design* measure requires major changes in hardware and software and is usually applied for new equipment. The *technology management* is a strategic measure to implement emerging technologies. (Böhner, Kübler & Steinhilper, 2013)

The definition of parallel principles allows to consider and apply the energy efficiency principles besides each other. These contain (Abele & Beckmann, 2012, p. 262; Erlach, 2013, pp. 346 f.; Junge, 2007, p. 10; Müller & Löffler, 2009, pp. 123 ff.; Ott & Cramer, 2009, p. 1019; Sauer et al., 2013, p. 35; Verl et al., 2011, p. 347):

- maintenance,

- machine control,

- machine load and dimensioning,

- reduce stand-by consumption,

- equipment efficiency,

- production planning and control,

- reduce process losses,

- energy reuse and recovery,

- production technology,

- optimize process chain, and

- substitute energy sources.

Maintenance is a business process to secure functionality of technical equipment and comprises inspections, service, and repair. Equipment wear may lead to a higher energy consumption, such as an increasing friction due to gear wear or a polluted filter in an air ventilator (Abele & Beckmann, 2012, p. 262). Hence, an appropriate maintenance strategy may increase energy efficiency.

Measures in the area of *machine control* mean to assign an appropriate operational state to the current requirements of a process. For example, peripheral components of a machine tool may be coupled to the manufacturing process, which means that exhaust air systems only need to be operating while the production is running (Müller et al., 2009, p. 125). Another example is the use of frequency-controlled pumps in order to adjust the conveyed volume to the current demand (Abele & Beckmann, 2012, p. 262).

A correct *dimensioning* is important in terms of energy efficiency since technical equipment usually operates less efficient in partial load (Müller et al., 2009, p. 125). Therefore, the dimension needs to be adjusted to the requirements of the particular application.

Reducing the *stand-by consumption* is relevant since equipment may consume a considerable amount of energy during non-productive time periods (Erlach, 2013, p. 346). This means either to provide an energy-saving mode for equipment (during planning) or to switch off equipment completely during non-productive times (during equipment operation).

The *equipment efficiency* is an important indicator to assess energy efficiency that is understood as the ratio between useful energy and total energy provided (Müller et al., 2009, p. 76). It refers to both production equipment (*e.g.*, electric motor in a robot) and process technology (*e.g.*, electrical transformers).

Production planning and control includes defining the production program, which among others determines the order of production jobs. Summarizing jobs may help to optimize the equipment load and, thereby, reduce energy consumption (Abele & Beckmann, 2012, p. 262).

Energy losses include energy output from a system that is not used for the process, such as excess heat or warm waste water. Although the reduction of energy losses is already integrated into the equipment or process efficiency, they may additionally occur for media supply systems (*e.g.*, leakage in compressed air systems), for which usually no efficiency is calculated (Müller et al., 2009, p. 127).

Excess energy may, otherwise, be used for *recovery and reuse*, such as excess heat that heats fresh water by using a heat exchanger (Müller et al., 2009, p. 128). Energy recovery may also be applied to use energy multiple times for the same process, *e.g.*, a motor that recovers energy in an intermediate circuit (Abele & Beckmann, 2012, p. 262).

The choice of production process and *production technology* highly influences the later energy consumption, such as the different efficiencies between hot forming and cold forming (Müller et al., 2009, p. 126) or between wet processing and dry processing (Abele & Beckmann, 2012, p. 262). Furthermore, the combination of various production processes in a *process chain* needs to be scrutinized. A typical approach is to integrate several processes, such as process-integrated quality control (Müller et al., 2009, p. 126) or hardening during grinding (Abele & Beckmann, 2012, p. 262).

Finally, the *energy sources* that represent the input to a production process may be *substituted*. This particularly influences the primary energy consumption due to different primary energy factors (*e.g.*, heating gas compared to electricity for heat generation). Another example is to substitute compressed air by electricity (Bayerisches Landesamt für Umwelt, 2009, p. 18).

While the aforementioned principles focus on the efficient usage of energy, MÜLLER ET AL. give a comprehensive understanding of the factory including energy generation, conversion, distribution, and storage (Müller et al., 2014, p. 218). This leads to a more general picture of energy efficiency principles (Figure 17).

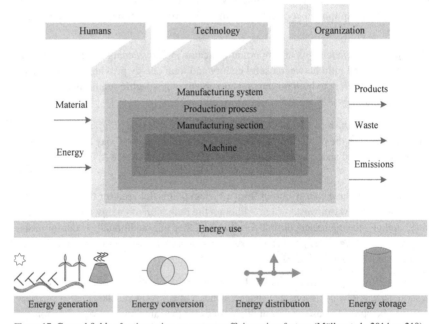

Figure 17: General fields of action to increase energy efficiency in a factory (Müller et al., 2014, p. 219)

The upper part depicts an overview of principles to reduce energy consumption, such as substituting energy carriers, increasing machine efficiency, and considering interrelationships between equipment in a manufacturing section. The measures to reduce energy usage refer to different levels: On the level of a single *machine*, the focus is on efficiencies and dimensioning. The *manufacturing section* considers the interrelationships between several machines, such as grouping equipment depending on its media requirements. The *production process* considers the entire chain of production steps (*e.g.*, integrate several machining processes into one forming step). The *manufacturing system* contains the production environment including building and building services.[3] (Müller et al., 2014, p. 219)

The lower part of Figure 17 demonstrates that actions to reduce energy consumption need to address energy generation, conversion, distribution, and storage as well (*e.g.*, use renewable energy, efficient compressors, avoid losses in pipes). (Müller et al., 2014, p. 219)

Energy efficiency principles provide a proficient guideline for enterprises in order to think holistically when increasing their energy efficiency. However, they are formulated in a general way, which may pose a barrier for the transfer to a concrete practical implementation. Therefore, they may be more helpful as a consulting instrument or to deduce a general strategy.

3.5.4 Systematic Methods for Energy Efficiency Improvement

In contrast to the techniques and instruments explained so far, systematic methods are characterized by a step-by-step procedure to guide a user to the identification of energy efficiency improvement measures. Methods for improving energy efficiency may be distinguished into analysis, assessment, and optimization methods (Despeisse et al., 2012, p. 33): *Analysis methods* are used to increase transparency on a system's energy consumption. The goal is to prioritize the subsystems as starting point for further optimization measures. *Assessment methods* also aim at creating transparency but extend their goal beyond a mere analysis. The goal is to compare concepts, *e.g.*, production technologies, regarding their effects on energy efficiency or environmental objectives. *Methods for the optimization* of production systems require an analysis but also describe how improvement measures are identified. Since the thesis focuses on the identification of improvement opportunities, the latter ones are focused in this section. The following sections describe the state of the art on methods to increase energy efficiency in both factory planning and factory management.

Integrating Energy Efficiency in Factory Planning Processes

ENGELMANN develops an approach to integrate energy efficiency as an objective into the factory planning process (Engelmann, 2009). This approach is described in more detail by E. MÜLLER ET AL. who develop the energy efficiency-oriented factory planning process (Müller et al., 2009, pp. 212 ff.). The method shows how to consider energy efficiency in factory planning activities (see Section 3.4.3). The identification of energy efficiency measures is supported by guiding principles for energy-efficient systems (see

---

[3] This understanding integrates all other aspects beside the production itself and, hence, considers the term manufacturing system as synonym to the factory. See Section 3.4.1 for the differentiation of these terms.

Section 3.5.3). Hence, the concept describes the points in the planning process, at which energy efficiency needs to be considered.

WIENDAHL ET AL. conceive the synergetic factory planning process that integrates the different perspectives of process planning and building planning (Wiendahl, Reichardt & Nyhuis, 2014). This approach aims at improving the sustainability of a factory by considering both the requirements of product and building. Hence, the approach contrasts the traditional planning approaches with successive stages, in which products and processes define the need for the building system. This meta-concept may be used to increase energy efficiency during factory planning but does not suggest concrete measures to do so.

F. MÜLLER ET AL. develop a concept for green factory planning that contains methods and tools in order to support the "green" objective in planning processes (Müller et al., 2013c). This concept aims at generating green planning modules to support planning tasks. For example, the layout planning module needs to consider the interaction between production and building systems (e.g., compressed air supply requirements). However, the concept focuses on a general level and does not describe a procedure to identify action approaches during factory planning.

CHEN ET AL. conceive an approach to integrate sustainability into the factory planning process (Chen et al., 2012). It is based on a model that represents the relation between factory system elements and sustainability aspects. The purpose is to designate the impact of factory planning decisions on sustainability objectives (e.g., the effect of machines on air quality). Transparency on these relations forms a starting point for improvements but the approach does not contain methodical support for this step.

DOMBROWSKI and RIECHEL develop a technique to evaluate the sustainability of a factory (Dombrowski & Riechel, 2013). This instrument considers sources and sinks in terms of sustainability. For example, a source is the water supply from a near river, whereas the emissions into the surrounding air represent a sink. The results of the assessment highlight the areas that should be considered in more detail. However, the concept does not support the identification of sustainability measures.

A method to forecast the energy consumption of a production system is described by WEINERT (Weinert, 2010): This approach uses building blocks that represent the power load of systems in different operational states. It is based on the assumption that the energy consumption of any process may be predicted by combining and parameterizing these building blocks. While the method initially focused on electricity, its extension allows considering several energy carriers, such as welding gases (Mose & Weinert, 2015). However, the initial data acquisition for creating the building blocks requires measuring the energy consumption of the equipment.

HAAG suggests an integrated methodology for planning, assessing, and optimizing the energy consumption in production systems (Haag, 2013). The method is based on a quantitative model of the systems within a factory. It includes production and peripheral equipment with their corresponding operational states. The operational states are characterized by their energy consumption and the time that a system spends in this operational state. Data sources may cover field data, expert knowledge, and standardized times taken from work planning (Haag, 2013, p. 74). The model allows an assessment of the current situation by applying various indicators (*e.g.*, specific energy consumption). Afterwards, an iterative optimization is conducted by studying the effects of parameter changes and developing different scenarios. This is supported by a simulation in order to consider the interrelationships among parameters and resources.

HOPF develops a methodology for modeling factory systems with regard to the objectives energy and resource efficiency (Hopf, 2016). After defining the purpose of the study, a qualitative model is generated that describes the function, structure, and hierarchy of a factory system. Subsequently, an optional quantitative model may be used in order to calculate key figures for a detailed assessment. The approach considers the entire factory system including production, building, and supply and disposal. The main goal is to describe factory systems holistically in early conceptual planning phases. The identification of improvement measures is supported by an overview of general action approaches but not focused within the method.

FRESNER ET AL. introduce an approach to identify improvement opportunities on a generic level by applying the Theory of Inventive Problem Solving (TRIZ) (Fresner et al., 2012). The TRIZ theory, as described below, is based on the analysis of a large number of patents and their corresponding functional principles (Altshuller, 1984). The identified 40 generic guiding principles can be applied to other inventive or product design tasks. Furthermore, the theory includes paradigms that support the creative thinking process, for example the basic concept of contradictions that need to be solved in a design process (Orloff, 2012).

The methodical approach of TRIZ in the context of cleaner production suggests a four-step approach (Fresner et al., 2012): At first, the current situation is analyzed including its useful and harmful functions. The next step is to develop the ideal final result following the TRIZ methodology. The ideal solution would provide the useful functions by eliminating the harmful functions. The optimization procedure is based on the identification of barriers between the current and the ideal situation. Creative methods, such as brainstorming, are carried out in an interdisciplinary team in this step. Finally, data on energy and material consumption as well as waste are collected in order to assess the feasibility of the identified options. This method provides a generic approach to create solutions for increasing energy efficiency. It does not suggest any concrete measures but relies on the creative thinking process of planning participants. While this allows a wide range of possible solutions, the success depends on the experience of the team members with TRIZ methodology and energy efficiency optimization.

A similar method is suggested by KUBOTA and DA ROSA (Kubota & da Rosa, 2013). They explicitly state the need to acquire quantitative data on the environmental effects of harmful functions. This may be achieved by gathering flowcharts within a company and conducting semi-structured interviews. In contrast to the aforementioned approach, they do not define an iterative approach. Instead, solution proposals are directly generated by applying the general TRIZ principles to the specific situation. The approach is supposed to be applied to a selected area rather than to the entire factory. The selection of relevant processes may be supported by estimates on environmental losses.

Increasing Energy Efficiency during Factory Management

ERLACH and WESTKÄMPER establish the concept of the energy value stream method (Erlach & Westkämper, 2009). Its goal is to create transparency on the energy consumption of a process chain and to develop improvement measures. The steps of the method include the energy value stream analysis, energy value stream design, and energy management. The analysis acquires the material and information flow for the production processes of the value chain. The energy efficiency is evaluated by calculating the energy intensity for each process, which is defined as the energy consumption to manufacture one product (Erlach & Westkämper, 2009, p. 32). These values are compared with reference values that document the state of the art. Based on this comparison, processes may be selected for further improvement. Afterwards, the value stream design aims at improving the efficiency of the value chain.

The energy value stream method is extended by integrating the objective to reduce carbon dioxide emissions (Erlach & Sheehan, 2014). The result is a visualization of the energy consumption in a value stream map. Furthermore, energy efficiency measures are identified and presented in an implementation plan. However, the acquisition of data for reference values to assess the energy intensity of a process poses a challenge. Although there are some initiatives on collecting and providing reference values for the energy consumption of manufacturing processes, these struggle to consider the variety of influential variables on energy consumption. Hence, only a wide range of reference values is available (Duflou et al., 2012a, p. 65), which makes it hard to use these values as basis for a comparison.

A similar methodological approach is suggested by REINHART ET AL., the Energy Value Stream (EVS) (Reinhart et al., 2010; Reinhart et al., 2011). The basic steps are measurement, visualization, analysis, generation of optimization measures, prioritization, and identification of interactions. Different modules are defined to visualize the processes as value stream (*e.g.*, production modules, transport modules, distribution modules). The analysis leads to the assessment of processes by applying performance indicators, such as the energy consumption per manufactured part. The energy value stream design contains the identification of optimization measures. This is supported by guiding questions, such as "Are the process parameters correct?". Therefore, the method creates transparency that allows for investigations into measures that could be taken to increase energy efficiency.

Another value stream based approach is developed by BOGDANSKI ET AL. (Bogdanski et al., 2013). The goal of this method is to assess an existing manufacturing system and to point out the influences of product design. The relevant data is acquired by power measurements. The results are presented in form of a value stream map that contains both production processes and building services. The energy consumption of the technical building services is assigned to the products that are produced during the considered period of time. Furthermore, the influence of product properties on the energy consumption is estimated. The authors do not explicitly provide an explanation about the identification of energy efficiency measures. However, the method creates transparency on the energy consumption in different operational states and the influence of product design. Therefore, the results can be interpreted by both manufacturing engineers and product designers which enables the identification of improvement measures.

Another extension of the value stream method is given by FISCHER ET AL., who integrate the selection of improvement measures (Fischer, Weinert & Herrmann, 2015). The so-called solution elements may contain technical measures as well as methodological approaches and frameworks. Within this approach, the value stream method is used in order to identify quantifiable energy and cost drivers, *i.e.*, parameters that affect the energy consumption and/or energy costs. Furthermore, design parameters are defined which represent main levers for increasing energy efficiency (*e.g.*, building layout). A procedure is described that provides a mapping between cost drivers, design parameters, and solution elements. By using this approach, solution elements may be selected by a partially automated algorithm based on the results of the energy value stream analysis.

STOCK describes a value stream approach to clearly distinguish between value-adding and non-value-adding energy consumption (Stock, 2016). The basis is to consider a manufacturing process while producing a part and to compare this with a so-called "air cut", *i.e.*, running the machining program without a part. The comparison is done by analyzing the load profile for both alternatives. Based on that, both the cycle time and the energy consumption can be distinguished into a value-adding and a non-value-adding share. In this understanding, value to the part is only added for the operations that the customer pays for (*e.g.*, machining is value-adding, whereas tool changes are non-value-adding). The approach provides transparency on the energy consumption of a process but the identification of energy efficiency measures is limited to the opportunities that are inferred from the load profile.

KRAUSE ET AL. propose a method based on the modeling of material and energy flows (Krause et al., 2012). Although the publication is focused on aluminum foundries, the approach is described in a general manner in order to apply it to different sectors. After a general system definition, the material, energy, and resource flows through the factory are measured and modeled. The level of detail for the necessary data can be adjusted to the goal of the study: In the presented case study, load profiles of some high prioritized machines are acquired, which allow to assign the energy consumption to operational states of these machines. Based on the calculation of indicators, the most promising fields of actions are identified. For these fields, improvement measures are identified and evaluated in terms of energy savings and carbon dioxide emission savings. The method provides a structured approach to identify "energetic hotspots" in a factory. However, there is no further guidance provided on how to deduce improvement measures.

Another approach that focuses on a factory model of material, energy, and waste flows is developed by SMITH and BALL (Smith & Ball, 2012). Its goal is a systematic identification of improvement opportunities within factories. The necessary information mainly covers the building geometry, layout, processes, and metered data on resource, energy, and mass balances. The approach integrates both production processes and technical building services. The first step is to define the settings, system boundaries, and targets. Afterwards, the so-called "factory gate analysis" collects available data from the utility suppliers (e.g., power supply company). This allows to identify large energy consumers within the factory. The next step is to develop a qualitative model of the building and the processes, i.e., building geometry, list of processes, and equipment. Next, the quantitative model is developed that contains information on the resource and energy consumption. Finally, an optimized process model is created by means of a simulation of the system performance. Depending on the level of detail of the simulation, local or even system-wide improvement opportunities can be analyzed.

BÖHNER describes an approach to identify and evaluate measures to save electricity within industrial enterprises (Böhner, 2013). The first step is to prioritize machines based on their nominal power, runtime, and expected savings, whereas the latter ones are identified through expert interviews. Furthermore, general information, such as the shift system, number of products (total and referred to each process), and the energy costs are acquired. In the next step, detailed power measurements are conducted for the highest-ranked machines. The acquired load profiles are used for the identification of measures. This is guided by the developed categories of energy efficiency measures (see "stairs of innovation" in Section 3.5.3). Later on, the measures are assessed according to their costs, saving potential, and transferability to other areas within the company. The measures are prioritized in a portfolio visualization with the two dimensions costs and benefits. The focus of the method is machinery but it may be extended to other systems within the factory.

THIEDE ET AL. describe a method that is focused on the application in SMEs (Thiede, Bogdanski & Herrmann, 2012; Thiede, Posselt & Herrmann, 2013). Its goal is to support the systematic identification of improvement potentials on energy and resource efficiency. At first, a list of all machines including their nominal power and operating time is developed. This is used for a visualization in a portfolio to deduce the further strategies: For machines with high nominal power, detailed measurements are suggested in order to understand the load profile. Machines with a low power level and high operating time do not require detailed measurements but might be suitable for measures with low investments. Furthermore, a regular update of the machine list is suggested in order to support the continuous improvement process. While the portfolio visualization helps to focus on prioritized machines, the identification of improvement measures is not supported.

WOLFF ET AL. describe the use of discrete-event simulation methods to forecast the energy consumption of manufacturing systems (Wolff, Kulus & Dreher, 2012). The goal of this approach is to evaluate the effect of changes in model parameters, *e.g.*, cycle times, on a system's energy consumption. By this, improvement potentials may be deduced, validated, and quantified before the system is realized. As input information, it is required to determine operational states of a machine and the power load within these states. Afterwards, the energy consumption is calculated depending on the time, during which a machine is in the respective operational state. Usually, simulation methods consider both material and energy flows. Data sources for the energy consumption or power load may be estimates from the nominal power, laboratory values from machine manufacturers or may be based on measurements that are conducted during the machine installation.

SCHACHT and MANTWILL describe a simulation method to support the planning processes of machinery and equipment (Schacht & Mantwill, 2012). The outcomes of the simulation are used to improve machine technology and machine operation. Furthermore, the results are used to dimension the technical building services.

DIETMAIR ET AL. simulate the energy consumption of machine tools by modeling their components (Dietmair, Verl & Wosnik, 2008; Dietmair, Verl & Eberspächer, 2011). The purpose is to validate the effect of energy efficiency measures after their implementation. Categories for measures are suggested and explained by means of a case study, including the selection of machine parameters, the programming of production machines, the production planning and control, and the optimization of numerical control programs.

ABELE ET AL. explain another simulation approach that is focused on machine tools (Abele, Eisele & Schrems, 2012). The purpose of this method is to estimate savings that may be achieved by applying improvement measures prior to implementation. The method is based on quantitative models that contain the energy demand of various components of a machine tool. The input data only requires data sheets but no measurements.

POHL ET AL. present a simulation approach to identify and evaluate optimization measures in compressed air networks (Pohl, Schevalje & Hesselbach, 2013). The purpose is to assess the interrelationships between the components of a compressed air system when implementing improvement measures. The input is a plan of the compressed air network including pipe lengths and diameters as well as information on the cycle times of the compressed air applications. As a result of the simulation, the economic profitability of measures is assessed.

RODRIGUEZ ET AL. use a combination of material and energy flow analysis and best available techniques (BAT) to identify energy efficiency improvement potentials (Rodríguez et al., 2011). The first step of the method is to quantify the material and energy flows within a factory. The flows with the highest consumption values are considered as important flows for the improvement. For these processes, BATs are identified. The documentation of BATs is done by the European Intergovernmental Panel on Climate Change (IPCC) Bureau (European Commission – Institute for Prospective Technological Studies, IPTS, 2016).

BUSCHMANN develops an approach to connect measured energy data with the knowledge of an industrial enterprise in order to support a continuous energy efficiency improvement (Buschmann, 2013). It demonstrates the efficient use of energy data that is acquired on machine level by means of an energy monitoring system. After the acquisition, the data is interpreted and compared with reference values. Based on that, potentials for improvement are identified. For the subsequent optimization, the existence of a company-wide database of possible measures is assumed. The final steps of the method are to evaluate the feasibility of a measure and to support the implementation.

3.5.5 Assessment of Existing Methods

The previous sections present an overview of methods and tools to increase energy efficiency in factories. There are several approaches to identify improvement measures. Norms and standards can be used for a high-level, strategic implementation of energy management but they do not provide operative solutions to identify improvement measures (Section 3.5.1). A long-term implementation of energy efficiency in a company requires to build up knowledge in the own organization instead of the mere relying on commercial offers and consultancies (Section 3.5.2). Energy efficiency principles may serve as a general checklist or consulting instrument, but still require effort to be transferred into a concrete application (Section 3.5.3). Therefore, methodical approaches are the most important category towards the goal of this thesis (Section 3.5.4). The criteria to assess the methods result from the current barriers for implementing energy efficiency and the applicability in factory planning (see Table 4). In the following paragraphs, the criteria are explained in detail, before the assessment of relevant approaches is conducted.

Table 4: Assessment criteria for existing methods to increase energy efficiency in factories

Criterion	Reason	Reference
Practicability of data acquisition	Lack of time and know-how	2.3
Systematic optimization procedure	Deduce measures after analysis	2.3
Completeness of objects	Variety of objects in factory systems	3.5.3
Completeness of energy flows	Structure of energy consumption in industry	2.2.4
Range of measures	Variety of possible measures	3.5.3
Specificity of measures	Transfer to concrete application	3.5.2
Applicability in factory planning	Factory planning tasks and approaches	3.4.3
Transferability	Lack of time and know-how	2.3

Practicability of Data Acquisition

The amount of data that is required to apply a method and the resulting effort, poses a barrier to industrial application. The hindering effect increases if the expected savings can hardly be quantified. Special skills might be required in order to collect the necessary data (*e.g.*, conducting measurements of electricity consumption or expert estimates on expected savings). This is a challenge especially for SMEs since they rarely have staff available for energy efficiency tasks. Many methods require to collect data on the energy consumption of equipment. Measurement tasks are characterized by a high effort for data acquisition (Steinhilper et al., 2012, p. 342). However, some approaches try to reduce this effort by preselecting equipment to be measured.

The practicability of data acquisition is considered low when detailed information on energy consumption and power load are necessary, *i.e.*, a load profile that depicts the power need over time, since this may only be acquired during detailed measurements. A medium practicability refers to aggregated quantitative information, such as an average power demand during operation, which may be found in data sheets from the equipment manufacturer. A high practicable method requires only qualitative information, such as a description of the energy flows that are required to operate a system.

Systematic Optimization Procedure

The presentation of methodical approaches in Section 3.5.4 demonstrates that many methods focus on the analysis and assessment of (partial) factory systems. This is used to create transparency and to prioritize areas of a factory for a subsequent detailed evaluation. The next step needs to be a consequent optimization approach. The interpretation of analysis results and the consequent suggestion of suitable measures requires experience in energy efficiency projects. Since this might not be available for the respective planning participants, it is important that the method contains a systematic description of how to identify measures.

An optimization procedure is considered little systematic when it only creates transparency on the energy consumption of a system without supporting the identification of improvement measures, *i.e.*, the user needs to interpret this information mainly independent. A medium assessment requires a checklist for improvement measures, such as guidance by a few energy efficiency principles. If a method provides a step-by-step procedure to identify improvement measures, this is understood as a highly systematic optimization.

Completeness of Objects

Section 3.5.2 discusses the necessity to consider a factory in its entirety in order to maximize the saving effect of energy efficiency strategies. This means that the building, production processes, logistics, process technology, and building services need to be considered. Depending on the industrial sector, the share of energy costs caused by these consumers may vary. However, depending on the origin and goal of a method, their application purpose may be focused on specific parts of a factory (*e.g.*, manufacturing processes).

A low assessment is given when a method only considers production processes, whereas a medium assessment requires to integrate technical building services. A high assessment may only be achieved when the entirety of factory systems is considered as described above.

Completeness of Energy Flows

Similarly to the aforementioned criterion, it is important to cover the entirety of energy flows within a factory. While electricity is often focused when analyzing energy efficiency, the main used energy carrier in industry is heat (see Section 2.2.4). Again, the shares highly depend on the industrial sector and need to be considered individually for each factory.

This criterion is assessed low for only one energy flow, medium for more than one energy flow and high if there is no limitation on the consideration of energy flows.

Range of Measures

Another assessment criterion is the limitation of the solution space, *i.e.*, the range of possible measures that are suggested by an optimization method. Similar to the variety of factory systems, there is a huge variety of energy efficiency measures in general. If they are limited due to the method's structure, improvement potentials of a specific case might not be exploited completely.

A low assessment refers to the case where measures are limited to specific model parameters (*e.g.*, time spent in an operational state). Many methods give supporting guidelines in form of a limited number of energy efficiency principles, which refers to a medium assessment. A high range of measures is provided if there is no general limitation.

Specificity of Measures

Time and know-how are limited resources in enterprises, which is why methods to improve energy efficiency should provide a comprehensive support. This includes the specificity of the resulting measures. The measures should, at best, refer to the considered system in order to simplify the implementation. That means if only general measures are suggested (*e.g.*, increase efficiency), there is still a lot of effort needed to transfer the measure to the concrete application. However, this may in turn require specific information about the system. Hence, it should be noted that there is a trade-off between required information and the specificity of results.

The criterion is assessed low when measures are described on a very general level. A medium assessment refers to methods that provide general measures which are focused on an application area. Considering a reference to the analyzed system results in a high assessment.

Applicability in Factory Planning

Numerous approaches support the identification of improvement measures during factory management. However, the highest influence on a factory's energy consumption can be achieved in early planning phases. Hence, it is important to integrate energy efficiency into factory planning processes. While this necessity is common sense in literature, many existing approaches in factory planning provide concepts on a general meta-level instead of supporting the identification of measures in detail. The fulfillment of this criterion may be limited by the required information (*e.g.*, scheduled time for a manufacturing step) and/or by the content of the tasks that are supported (*e.g.*, sequence of production orders).

A low assessment is given when a method is only applicable in factory management. If some selected planning aspects (*e.g.*, work planning) are supported, this leads to a medium assessment. Methodical support throughout the entire factory planning process refers to a high assessment.

Transferability

Finally, the methods are assessed regarding the clarity of their description. A transferable method is characterized by a clear and detailed description of each step, which allows persons with professional knowledge in the respective area to apply the method. This is an important criterion in order to ensure the applicability in industrial applications. Difficult calculations, sophisticated information and communication tools, and special methodical skills might reduce the transferability of an approach. Moreover, due to the origin of a method, the steps of the procedure might not be explained on a high level of detail. For example, approaches that are applied in consulting are not completely transparent in order to maintain a competitive advantage.

A method is considered as little transferable if there is a lack of a clear methodical description. When the steps are completely explained but require specialist expert knowledge in the area of energy-efficient production, a medium transferability is assessed. A high assessment requires a clear, detailed, and understandable documentation of each step in the procedure.

Conducting the Assessment

The most promising methods from the presentation in Section 3.5.4 are analyzed according to the defined assessment criteria. Therefore, a pre-selection is performed with regard to the following aspects: In the area of factory planning, there are several concepts that point out the importance of energy efficiency.

These meta-concepts describe general guidelines for the factory planning process but do not support the identification of energy efficiency measures. Other approaches are limited to the evaluation of specific elements in a factory, such as machine tools or compressed air systems. Although these may represent an important part of factory systems, the detailed consideration of a specific object system does not fulfill the purpose of this thesis. Hence, these two groups are excluded from the assessment. The assessment results for the selected approaches are depicted in Table 5.

It can be seen that none of the approaches fulfills all of the requirements. The majority of the approaches follows the scheme of a quantitative analysis in order to create transparency and to evaluate the energy consumption. The equipment or areas that have a higher energy consumption are assumed to have a high saving potential and should be focused on. The need to acquire energy consumption data hinders the application of methods in early planning phases. Additionally, this approach puts an emphasis on the analysis rather than the improvement of a system.

Other approaches that focus on the improvement rather than the analysis lack a methodical description of that step. They are based on creative techniques and largely depend on the knowledge and experience of the planning participants.

Furthermore, a focus on manufacturing processes and equipment can be observed. A systematic optimization method should, however, focus on the entire factory system including building, technical building services, and peripheral processes (*e.g.*, logistics). Several methods require the use of specific techniques (*e.g.*, simulation models) although the corresponding knowledge might not be available in industrial companies, especially in SMEs.

Table 5: Assessment of existing methods to increase energy efficiency of factory systems

	Practicability of data acquisition	Systematic optimization procedure	Completeness of objects	Completeness of energy flows	Range of measures	Specificity of measures	Applicability in factory planning	Transferability
Bogdanski et al., 2013	◇	◇	◈	◇	◈	◆	◇	◆
Böhner, 2013	◇	◈	◇	◇	◈	◈	◇	◆
Buschmann, 2013	◇	◆	◈	◆	◈	◆	◇	◆
Engelmann, 2009	◆	◈	◆	◆	◈	◇	◆	◆
Erlach & Westkämper, 2009	◇	◈	◇	◇	◈	◆	◇	◈
Fischer, Weinert & Herrmann, 2015	◇	◆	◈	◇	◆	◈	◇	◆
Fresner et al., 2012	◆	◈	◆	◆	◆	◈	◆	◇
Haag, 2013	◈	◆	◈	◇	◇	◆	◈	◆
Hopf, 2016	◆	◈	◆	◆	◈	◇	◆	◆
Krause et al., 2012	◇	◇	◈	◆	◈	◆	◇	◆
Kubota & da Rosa, 2013	◇	◈	◆	◆	◆	◈	◆	◇
Reinhart et al., 2010	◇	◈	◈	◆	◈	◆	◇	◆
Rodríguez et al., 2011	◈	◆	◇	◆	◆	◆	◈	◇
Smith & Ball, 2012	◈	◆	◈	◆	◆	◈	◇	◈
Stock, 2016	◇	◇	◈	◈	◇	◈	◇	◆
Thiede, Bogdanski & Herrmann, 2012	◈	◇	◈	◇	◈	◈	◇	◆
Weinert, 2010	◇	◇	◇	◆	◇	◈	◈	◆
Wolff, Kulus & Dreher, 2012	◈	◇	◇	◇	◇	◆	◈	◆

Legend: ◇ low   ◈ medium   ◆ high

As a summary for the assessment, there is a lack of a methodical approach that provides a structured improvement of the entire factory system throughout the factory life cycle, *i.e.*, which may be applied in early planning phases. The resulting suggestions should not be limited to a-priori defined specific measures. The approach needs to be well-defined and based on solutions that are easy to implement, as far as possible, in order to support the implementation in industrial enterprises.

3.6 Interim Conclusion on the State of the Art

Modeling is the process to describe a specific extract of reality in a simplified and goal-oriented manner. Models can be used for description, explanatory, forecast, decision, optimization, and simulation purposes. Depending on the modeling purpose, various types may be applied and various aspects of the original system (*e.g.*, structure, behavior) may be emphasized. The modeling procedure is guided by general principles and the resulting quality of a model can be assessed by the principles of proper modeling.

System theory is a general approach to describe and understand diverse objects. A system is composed of several elements that are related to each other and collectively fulfill a specific purpose. Systems engineering is the concept to apply systems thinking to engineered systems. Besides the mental framework as part of the systems engineering philosophy, this contains the problem-solving paradigm as general procedure to resolve issues. Decision-making methods support the selection between alternative solution approaches.

Decisions in factory planning and management require information and knowledge on the systems that are to be planned. Knowledge representation has the task to visualize this knowledge and uses both graphic-based and matrix-based representation methods. Graphics provide an overview of concepts, structures, tasks, and activities and support interdisciplinary communication. Matrices describe knowledge in a more abstract form and are closer to the processing of knowledge.

Factories are complex socio-technical systems that are characterized by a variety of technical facilities and an organizational structure in order to manufacture a product with suitably divided labor. The behavior and interrelationships in factories may be analyzed and optimized by means of the systems theory.

A factory is integrated into an environment and connected to it by different flows, such as material, energy, and waste. The elements in a factory comprise the basic production factors equipment, material, and personnel. These elements are interlinked by structures and processes, which may be characterized as information, material, energy, capital, and personnel flows.

When modeling a technical system, different views may be pursued, such as the functional, structural, and hierarchical perspective. The functional perspective focuses on the main function of a system, *i.e.*, the generation of a system output from a system input, whereas the structural view considers the inner structure of a system that is used to achieve its functionality. A hierarchical view organizes systems in a relationship of subordination and allows to gradually distinguish systems (*e.g.*, consider production and peripheral equipment that belongs to a manufacturing area). These various modeling perspectives help to analyze a factory system holistically.

Factory planning is the systematic process to plan a factory and comprises a variety of tasks, such as decisions towards the factory location, building, production processes, peripheral processes, logistics, and staff planning. Models for the factory planning process help to structure the necessary activities in a procedure of planning phases and planning steps.

Increasing energy efficiency in factories is supported by a variety of norms, standards, instruments, and methods. Systematic methods that describe a procedure which may be implemented by an enterprise itself are especially important since they allow a long-term development of energy efficiency knowledge. The assessment of existing approaches shows that the majority focuses on a quantitative analysis in order to create transparency on the energy consumption rather than on the improvement with suitable measures. The effort that is required to acquire measurement data tends to be high and poses a barrier for implementation. This is especially true for projects taking place in early planning phases, in which this data might not be available. The analysis of the state of the art reveals that there is a lack of a methodical approach to identify solution approaches for increasing energy efficiency with reduced effort, i.e., through a purely qualitative description. Against this background, a qualitative method to identify energy efficiency measures for factory systems is developed in Chapter 4.

4 Method to Identify Energy Efficiency Measures for Factory Systems

Chapter four contains the development of the method to identify energy efficiency measures for factory systems. At first, goals and requirements are described as a result of the previous findings on implementation barriers and deficits of existing approaches. Based on that, the framework for the method is developed and the main research contribution is emphasized. Afterwards, the concepts for modeling the relevant domains are explained in detail. Finally, a procedure model is described that integrates the application of the separate concepts.

4.1 Goals and Requirements

The goal of this thesis is to develop a method to identify energy efficiency potentials for factory systems. Based on the findings from the analysis of the state of the art, the requirements for the methodical approach are described in the following paragraphs.

The high effort for data acquisition represents a barrier for the applicability of methods (see Section 2.3). This is especially true for approaches that require, for example, load profiles as input information (see Section 3.5.5). Hence, the methodical approach should pose *low data requirements*, which means that it is not based on measurements of the energy consumption. Instead, the requirements should be limited to qualitative information on the factory planning task.

A purposeful optimization of energy efficiency needs to address the *entire factory system* (see Sections 3.4.2 and 3.5.2). However, most of the existing approaches focus on production processes instead (see Section 3.5.5). Since the method addresses factory planning tasks, it needs to consider all relevant systems within factories, including manufacturing, assembly, process technology, building, and building services.

The method should be tied to the requirements of *factory planning tasks* (see Section 3.4.3). These are characterized by a high complexity, usually low detailed information and interdisciplinary project teams. Hence, a generic approach is developed that may be adjusted to the requirements of a specific use case. The availability of data refers to information on the energy consumption of equipment, which is hardly available during early planning phases (see Section 3.5.5). Connected to this criterion is the overall goal to reduce the data acquisition effort for applying the method. Finally, interdisciplinary teams result in different perspectives of each project participant, which need to be taken into account.

Many existing instruments support the analysis and assessment in order to create transparency on energy efficiency as a basis for deducing improvement opportunities. On the contrary, this method is supposed to provide solution approaches, *i.e.*, concrete measures to increase energy efficiency. Hence, the approach should focus on the *optimization phase* of an energy efficiency project.

Finally, a transparent and comprehensible method is required that allows a manageable transfer to industrial applications. This is due to the fact that a long-term improvement of energy efficiency is based on knowledge in a company instead of relying on external expert knowledge (*e.g.*, through consultancies). Moreover, finding relevant information for a specific use case and transferring this information requires high effort and time (see Section 3.5.2). Therefore, a *systematic step-by-step procedure* needs to be developed as part of the method to guide a user through the application.

As a result of these requirements, a new general approach for energy efficiency projects is realized (Krones & Müller, 2014a, pp. 506 f.). Due to the focus on qualitative information as methodical input, the steps of an efficiency project are reorganized. Based on a qualitative analysis of a planning situation, energy efficiency measures are suggested. Planning their realization requires estimates on costs and benefits in order to assess the feasibility. This assessment is based on general information since there is no case-specific quantitative data available in this step. If necessary, a quantitative analysis for a specific area may be conducted subsequently but is out of the scope of this thesis.

Figure 18 contrasts the main state of the art approach with the desired rough procedure of the methodical approach.

Figure 18: Schematic comparison between state of the art approach and suggested methodical approach for energy efficiency projects

The initial analysis is limited to qualitative information, which leads to a reduced effort as compared to a detailed quantitative analysis. The identification of measures is supported by a systematic procedure, whereas the state-of-the-art approach analyzes the acquired quantitative data and usually requires expert knowledge to deduce measures. The next step includes the assessment with general information, which means that information on an energy efficiency measure is provided in order to support the assessment. The following two steps are optional: A partial quantitative analysis may provide more detailed information on the system. It is believed that the effort for this additional step is still lower since a deeper understanding helps to limit the extent of measurements. If a quantitative analysis is conducted, the results need to be assessed in detail for this partial system. The final implementation step does not differ between the two project approaches.

It is expected that the overall effort for an energy efficiency project can be reduced through applying the method. This compensates for the lower level of detail of the results since subsequent detailed measurements and analyses are possible.

4.2 Overview on Methodical Approach

The method aims at supporting planning participants in increasing the energy efficiency of factory systems. Suitable energy efficiency knowledge should be provided in a purposeful way based on a qualitative analysis. Prior to qualitative modeling is the identification of domains that are necessary to fulfill the defined goals. The general task is to assign energy efficiency knowledge to a factory planning project, including energy efficiency measures and further information towards their assessment. The method is understood as a procedure to select and combine relevant information tailored to a specific use case. However, the informational content is part of a general, *i.e.*, not case-specific, knowledge base.

The description of the factory planning project needs to capture its socio-technical aspects, *i.e.*, both the technical object system and the actors that influence the factory system. Furthermore, the planning project is characterized by its conditions, *e.g.*, the respective planning stage. The interface between the planning project and the energy efficiency measures is designed by the so-called energy efficiency influential parameters, *i.e.*, parameters that are featured in the factory system and affected by energy efficiency measures. Hence, the model contains a qualitative description of parameters that represent a system's energy consumption without considering the concrete quantitative effect. Figure 19 depicts the methodical approach in form of a concept map (see Section 3.3.2).

The left part shows the two general areas of the factory planning task and the energy efficiency knowledge that are to be matched together. The *object system* represents the technical components of the factory system, *i.e.*, the factory building including production equipment and supply and disposal systems. It is characterized by a compilation of parameters that may influence its energy efficiency, the *energy efficiency influential parameters*.

The *actor* may be any person or group of persons that has a role towards the object system. This includes planning participants that are responsible for the projecting of equipment or persons that are affected from planning decisions, *i.e.*, the staff working in a factory. The *project characteristics* comprise "soft" aspects of both the factory planning project and the energy efficiency project. Examples are the factory planning phase, during which the project takes place, or the expected implementation time for a measure.

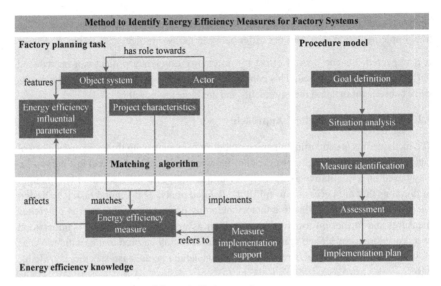

Figure 19: Overview on structure of the methodical approach

The *energy efficiency measures* represent short action approaches to reduce energy consumption and are linked with *measure implementation support* that provides details on a measure's realization. The assignment task is realized by a *matching algorithm* that is based on matrix-based knowledge representation (see Section 3.3.2).

The right part of Figure 19 shows the *procedure model* as part of the method. Initially, goals are defined for the method's application. Afterwards, the situation, *i.e.*, the planning task, is analyzed qualitatively. Based on this analysis, energy efficiency measures are identified and assessed with the help of the measure implementation support. Finally, an implementation plan is developed that may contain the realization of selected measures.

In the following sections, the modeling concepts for the domains object system, energy efficiency influential parameters, project characteristics, and actor are explained. This means that relevant aspects that need to be integrated into the description of the respective domain are discussed.[4] Afterwards, the structure for describing both the energy efficiency measures and the measure implementation support is developed. Based on the structure description of these two areas, the matching algorithm is generated. Finally, the procedure model for applying the method is developed.

---

[4] It should be noted that the method provides the general structure of how to describe each domain rather than determining each criterion specifically.

4.3 Description of the Object System

The purpose of this domain is to describe the characteristics of the technical system that should be improved by the efficiency project. The fundamentals of factory systems (see Section 3.4.2) are used to model the object system. The classifying criteria comprise general information on the entire factory and specific aspects of the systems in a factory.

Technical components in a factory are described by modeling hierarchy, function, and structure of a system (Hopf, 2016, p. 63; Ropohl, 2009, p. 75): The hierarchical concept highlights that a system is – on the one hand – composed by other subsystems and – on the other hand – part of a supersystem. The functional aspect considers a system as a "black box", *i.e.*, it describes the function of a system within its environment without taking the inner structure into consideration. Finally, the structural concept considers a system as a "white box" and, thereby, emphasizes the inner structure. The structural concept is discussed in more detail in Section 4.4 with reference to the parameters that influence a system's energy efficiency.

4.3.1 General Characteristics of Factory Types

Describing general characteristics of the factory is important in order to provide the first limitation on suitable energy efficiency measures. Factories may be classified into various types according to economic and technical-organizational characteristics (Wirth, Schenk & Müller, 2011, p. 799). The purpose of identifying factory types is to discuss similarities and differences between factories, which in turn need to be considered during factory planning. It is assumed that similar factories have similar energy saving potentials.

The *company size* is determined by the number of employees, turnover, and the annual balance sheet total. According to this definition, the category of micro, small and medium-sized enterprises (SMEs) comprises companies with less than 250 employees, an annual turnover not exceeding 50 million Euro, and/or an annual balance sheet total not exceeding 43 million Euro. (European Commission, 2003, p. 4)

As part of the organizational characteristics, a factory may be distinguished with regard to the manufacturing type, manufacturing form, and manufacturing principle. The *manufacturing type* describes the set of products: An individual production means to produce only one or a small number of identical products individually for a specified set of customer requirements (*e.g.*, special-purpose machines). In the series production, a number of defined products is manufactured continuously in small, medium, or large series (*e.g.*, automotive). Mass production comprises the manufacturing of a standardized product in a large amount (*e.g.*, cement). (Aggteleky, 1990, pp. 478 ff.)

The *manufacturing form* characterizes the organizational structure of production. A single spot production refers to manufacturing at a single place (*e.g.*, large diesel engines). In a job shop production, workshops are spatially grouped depending on their function (*e.g.*, turning shop, milling shop) and the product is processed on the different stations (*e.g.*, flanges, which are turned, drilled, and finished). A group production summarizes equipment and workstations depending on a product or component (*e.g.*, packaging cell at the end of a manufacturing line). The flow shop production arranges the equipment in the order of the manufacturing steps (*e.g.*, automotive assembly line). (Aggteleky, 1990, pp. 480 ff.)

The type of linking between manufacturing steps is characterized by the *manufacturing principle*. Linear, parallel, nest, asterisk, and workshop principles of manufacturing are distinguished depending on the direction of structures and processes between system elements. (Aggteleky, 1990, pp. 484 f.)

From a technical perspective, factories may be differentiated according to the degree of automation and the mainly applied technology. The *automation* may be either none, mainly manual, mainly automated, or fully automated (Kampker, Franzkoch & Hilchner, 2011, p. 570). The applied *technology* may be production technology for discrete parts manufacturing, process technology for bulk materials, or a combined technology (Wirth, Schenk & Müller, 2011, p. 799).

Besides these characteristics, factories are considered similar when belonging to the same industrial sector. Several classification schemes can be used for this task. In Germany, the Federal Statistical Office provides a classification of the manufacturing industry into 23 categories on the first level (Statistisches Bundesamt, 2007), which is depicted in Table 6. In the USA, the North American Industry Classification System (NAICS) is provided by the United States Census Bureau (United States Census Bureau, 2012).

4.3.2 Hierarchical Description of Factory Systems

The hierarchical description separates a factory system into vertical levels. It follows the concept of systems engineering to divide systems into the necessary level of detail (see Section 3.2). This means that a system may be subordinated to a supersystem or split into various subsystems.

Literature provides different possibilities to distinguish between hierarchical levels: The VDI guideline 5200 on factory planning suggests the levels work center, segment, building, plant, and production network (VDI 5200, Part 1, p. 7). MÜLLER ET AL. describe the hierarchical factory model as being composed of supply chain network, factory, building, production area, group of work centers, and work center (Müller et al., 2009, pp. 41 f.).

DUFLOU ET AL. further differentiate between the supply chain level and the multi-factory level in a comprehensive literature review on energy and resource efficient manufacturing (Duflou et al., 2012b, p. 588). Components and drives are considered as a more detailed level when explaining opportunities to increase energy efficiency (Günthner, Galka &

Tenerowicz, 2009, p. 11). With regard to energy management and energy monitoring tasks, the levels factory, department/process chain, and unit process/machine can be considered (Bogdanski et al., 2012, p. 541).

Table 6: Classification of manufacturing enterprises into industrial sectors (Statistisches Bundesamt, 2007)

No.	Category	Exemplary products
10	Food products	Meat, fish, oil, milk, pastries, sugar, coffee
11	Beverages	Beer, wine, mineral water, liquor
12	Tobacco products	Cigarettes, cigars
13	Textiles	Threads, filaments, woven material, technical textiles
14	Wearing apparel	Clothes, working clothes
15	Leather and related	Bags, coats, shoes
16	Wood and cork	Chipboard, veneer wood, hardwood floors
17	Pulp and paper	Pulp, paper, carton, cardboard, office supplies
18	Print industry	Newspapers, magazines, books
19	Coke and refined petroleum	Coke, petroleum
20	Chemicals	Industrial gas, dyes, soap, glue, synthetic fibers
21	Pharmaceutics	Medicine, vaccines
22	Rubber and plastics	Tires, plastic foil, plastic packaging
23	Non-metallic minerals	Glass, glass fibers, ceramic, tiles, bricks
24	Basic metal	Raw iron, metal pipes, wires, molded products
25	Fabricated metal	Screws, shafts, sheet metal, tools
26	Computers and electronics	Electronic components, circuit boards, measuring devices
27	Electrical equipment	Batteries, electric motors, lamps, household appliances
28	Machinery and equipment	Combustion engines, pumps, gears, machine tools
29	Motor vehicles and trailers	Cars, car bodies, accessories of motor vehicles
30	Other transport equipment	Ships, yachts, locomotives, airplanes, spacecrafts
31	Furniture	Office furniture, mattresses
32	Other	Jewelry, musical instruments, toys

It should be noted that the purpose of this thesis is focused on factory systems, which is why the production network is not considered here. Figure 20 shows the hierarchical model of a factory system. There, the highest level is represented by the factory, *i.e.*, the production site. The site may comprise various buildings, which in turn contain several divisions. The next hierarchical level is the segment that is composed of several work centers. Differentiating between segment and work center is conducted by analyzing the individual purpose, *i.e.*, a work center follows an individual purpose (*e.g.*, machine tool), whereas a segment contains work centers that are somehow interlinked. Objects may be summarized to segments and divisions on a spatial or functional basis (see Section 4.3.3).

Finally, a component is part of a work center that does not fulfill a purpose in terms of the production system by itself (*e.g.*, drive).

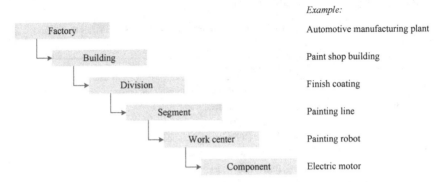

Example:

Automotive manufacturing plant

Paint shop building

Finish coating

Painting line

Painting robot

Electric motor

Figure 20: Hierarchical model of the factory system

4.3.3 Functional Description of Factory Systems

The functional perspective highlights the function of a system within its environment without considering the inner structure. Therefore, the functional description of a factory system describes the different tasks of a system, its input and output, and the operational states. Since a technical system is represented by its material, energy, and information flow, the tasks of a system may address an interaction with any of these flows.

Product functions can be classified based on different levels of detail. For example, production equipment, logistics, quality management, control, information, and communication systems are functional departments that fulfill production and production-relevant tasks (Schenk, Wirth & Müller, 2010, p. 5). Additionally, the required energy and media needs to be supplied and the work environment needs to be maintained through suitable building services. Furthermore, the building structure and its components are relevant parts of a factory.

From an energy efficiency perspective, the systems in a factory are functionally categorized into electromechanical drives, compressed air, lighting, information and control technology, process heat, process cold, heating and air ventilation, and building (Müller et al., 2009, pp. 159 ff.).

Depending on their function, factory systems can be classified into provision, generation, conversion, storage, transportation, usage (consumptive or non-consumptive), recovery, and emission with regard to material or energy flows (Hopf & Müller, 2013a, p. 138). A system may fulfill varying functions with regard to different objects. For example, a pneumatic compressor uses (consumes) electrical energy and generates compressed air.

A more detailed specification for the functional description is represented by the functional factory system structure (Figure 21): The factory system is distinguished by the production system, the building system, and the supply and disposal system. The production system contains the manufacturing, assembly, and logistics system, *i.e.*, the objects that directly relate to the product. The building system includes the structural parts of the building (*i.e.*, building shell, interior construction, supporting structure) as well as the property site and outdoor facilities. The supply and disposal systems contain process technology and building services, whereof the first category provides media for the production and the latter one supplies the building. (Hopf, 2016, pp. 93 ff.)

Factory		
Production system	Building system	Supply and disposal system
– Manufacturing: primary shaping, forming, cutting, joining, coating, changing material properties – Assembly and handling – Logistics: storage, transport, order picking, packaging	– Building structure – Property site – Outdoor facilities	– Building services: water and waste water, heat supply, ventilation, air-conditioning, electrical power, lighting, information technology, conveying systems, building automation – Process technology: use-specific systems

Figure 21: Functional description of a factory system (adapted from Hopf, 2016, p. 95)

The functions of the production system may be further specified. The manufacturing functions contain the processes primary shaping, forming, cutting, joining, coating, and changing of material properties (DIN 8580, p. 7). The assembly system is characterized by a variety of complex tasks that might be manual, automated, or semi-automated (Feldmann, Schöppner & Spur, 2014, p. 459). The logistics system in a factory fulfills the functions of storage, transport, order picking, handling, and packaging (Martin, 2014, p. 9).

The supply and disposal system contains the functions water and waste water, heat supply, ventilation and air conditioning, high-voltage electrical power (including switchgears), lighting, information technology, conveying systems, use-specific equipment (*e.g.*, cleaning), and building automation systems (DIN 276-1, pp. 16 ff.). The systems are assigned to building services or process technology depending on whether they supply the building system (*e.g.*, air conditioning) or the production system (*e.g.*, process heat).

Each technical system may be represented by a functional model. This model contains its primary function, the transformation process of energy, material, and information flows while considering the different operational states of the system (Hopf, 2016, pp. 68 f.).

The general functional model is graphically represented in Figure 22. The visualization corresponds to the method for Factory System Modeling in the context of energy and resource efficiency (FSM*ER*, Hopf, 2016).

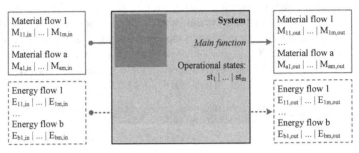

Figure 22: General functional model of the object system (adapted from Hopf, 2016, p. 69)

The interaction between the system and its environment is represented by the input and output of material and energy flows. The input material flow contains raw material, pre-products, and auxiliary material, while the output material flow contains products, by-products, and waste (Hopf, 2016, p. 68). The raw material and the pre-products are the basis for manufacturing a product, whereby the raw material is unprocessed. For example, bauxite is a raw material to produce aluminum, whereas an aluminum sheet is a pre-product to manufacture aluminum pipes. The auxiliary material is required for a process but does not enter the product (*e.g.*, lubricant).

On the side of the output, the product is the desired result of the process and by-products represent other physical results that may be used for a different purpose. Waste is the share of the output with no further utilization and needs to be recycled or disposed. The energy flow is also characterized by input (*e.g.*, electricity) and output that can occur in the form of waste heat, emissions, radiation, noise, and vibrations (VDI 4075, Part 1, p. 10).

The transformation process conducted by the system is described in the gray box of Figure 22. It includes selecting the primary function from the aforementioned classification (*e.g.*, manufacturing, material transport, energy generation). The lower section contains the operational states of the system (*e.g.*, operating, stand-by). The input and output are specified with regard to the operational states. For example, the flow $E_{1m,in}$ describes the energy input of item 1 in operational state m. (Hopf, 2016, pp. 68 ff.)

4.4 Description of the Energy Efficiency Influential Parameters

Since the goal of the method is to reduce the energy consumption of a system, a model is required to describe variables that influence the energy consumption.[5] This is understood as a qualitative description of parameters that influence the quantity of energy flows. In contrast to the hierarchical and functional system description as explained before, these parameters reflect the structural perspective on a system (see Section 4.3). The definition aims at energy flows and is not limited to energy consumption, *i.e.*, it may consider other functions as well, such as energy generation. It should be noted that the quantitative effect is not considered. In the following section, general basic approaches for describing influential parameters are explained. Afterwards, more details are discussed with regard to the hierarchical level of the respective object system.

4.4.1 Extended Functional Model of the Object System

Energy is represented by the integral of power over time:

$$W = \int_0^T P(t)\, dt, \text{ with } t \in [0, T].$$ (4.1)

Assuming a constant power need over time, this equation simplifies into the product of power and time. Hence, energy consumption depends on the power need of the elements within a system and the running time of the system:

$$W = P \cdot T.$$ (4.2)

More specifically, different operational states of a system can be distinguished depending on the performed task. Assuming that a system has n different operational states, the energy consumption can be calculated as

$$W = \sum_{i=1}^{n} P_i \cdot T_i.$$ (4.3)

The power consumption in each operational state depends on the composition of active and inactive elements (or subsystems) within the system. Therefore, an operational state of a system is characterized by the operational states of the subordinated systems and

---

[5] The term "energy efficiency" in general addresses the ratio between a useful output and energy input (see Section 2.1). The approaches in this thesis focus on the reduction of energy input. With regard to the scope of factory planning, the term "energy demand" would be more suitable since the relevant object system does not have to exist at that point of time. However, using the term "energy consumption" is commonly accepted in literature and will be applied throughout.

elements. Let m be the number of relevant elements within a system. If each element has only two operational states (on or off) with a power consumption only in the status "on", the equation changes into

$$W = \sum_{i=1}^{n} P_i \cdot T_i = \sum_{i=1}^{n} T_i \sum_{j=1}^{m} P_j \cdot \alpha_{i,j}, \qquad (4.4)$$

whereby the variable $\alpha_{i,j}$ indicates whether the element j is active in the operational state i. In general, the elements may have different operational states as well. In this case, each element is characterized by a power need in its individual operational state. Then, the energy consumption of the system is calculated as

$$W = \sum_{i=1}^{n} P_i \cdot T_i = \sum_{i=1}^{n} T_i \sum_{j=1}^{m} \sum_{k=1}^{p_m} P_{j,k} \cdot \beta_{i,j,k}, \qquad (4.5)$$

whereby $P_{j,k}$ contains the power consumption of element j in its individual state k and $\beta_{i,j,k}$ indicates whether the j-th element is in the individual state k if the system is in state i. The subsystems may influence each other, resulting in dependencies among their operational states. For example, an error in a machine tool may force it to stop the operation, which causes the cutting fluid pump to stop its operation as well. These dependencies need to be accounted for properly when defining the operational states of a system and its subsystems.

All of the introduced equations are based on the simplistic assumption that the power need is constant over time. Since this is usually not the case, the power need can be replaced by an average power demand in each operational state. Otherwise, functions may be modeled to represent the load profile, *i.e.*, the power need over time.

KRÖNERT ET AL. suggest the use of piecewise linear functions to model the function of power need (Krönert et al., 2013, p. 410). This means that the power need in an operational state is represented by a sequence of linear functions:

$$P_i(t) = \begin{cases} a_{i,1} \cdot t + b_{i,1} & \text{for } t \in [T_{1,1}, T_{1,2}], \\ \quad\vdots & \\ a_{i,\ell} \cdot t + b_{i,\ell} & \text{for } t \in [T_{\ell,1}, T_{\ell,2}]. \end{cases} \qquad (4.6)$$

The index ℓ describes the numeration of the pieces of the function and the coefficients $a_{i,\ell}$ and $b_{i,\ell}$ define the linear function within the boundaries. These functions may be defined for each operational state i.

Opposed to using linear approximations, WEINERT suggests polynomial functions to model the power demand in an operational state (Weinert, 2010, p. 75):

$$P_i(t) = \sum_{p=0}^{g} a_{i,p}(t - t_{i,0})^p, \qquad (4.7)$$

with a polynomial degree p between six and nine (Weinert, 2010, p. 75).

The previous findings imply to describe the energy consumption of a system with three aspects: First, the components of a system are determined. Secondly, the operational states describe the usage of these components to fulfill various functions. Finally, the information is combined to represent the load profile over time. In the case of considering the total energy consumption over a certain period of time, an integrative consideration is sufficient.

If the breakdown of energy consumption into power and time is not necessary, general energy performance indicators might be considered. These represent key figures that generally influence the energy consumption of a system. Energy key figures can be differentiated into absolute and relative indicators, whereas the latter ones put the energy consumption into relation with other information (Linke et al., 2013, p. 557). One example is the ratio between the energy consumption of a factory and the number of manufactured products. In this understanding, the energy consumption of a system is modeled by taking the most important influential indicators into account. In most cases, the model only includes one parameter. In the example above, the number of products is considered as the most important parameter for the energy consumption of a factory.

Subsuming the previous explanations, models to describe the energy consumption can be classified into object-oriented, state-oriented, and indicator-oriented approaches: Object-oriented approaches consider the different physical components within a system and the share of energy consumption that is caused by them. State-oriented approaches are based on the different operational states and their respective energy consumption. Indicator-oriented approaches describe a mathematical relation between influential parameters and the resulting energy consumption but do not distinguish between the influence of power need and time. Therefore, it can be deduced that the energy consumption of a technical system is mainly characterized by its operational states, the status of subsystems and elements within these states, and parameters that may influence the energy consumption (either within a specific operational state or related to the total energy consumption).

Based on this, the energy efficiency influential parameters are represented as an extension of the functional system model (see Figure 23 for an example). The functional model contains the operational states and may comprise relevant subsystems when a hierarchical structure is considered. The indicators that affect the energy consumption of a system are specified in the description of input and output. These parameters may either influence the energy consumption in total or with regard to a specific operational state.

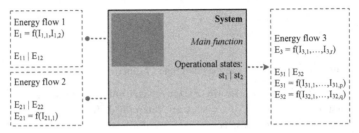

Figure 23: Extended functional model with energy efficiency influential parameters (example)

The exemplary system in Figure 23 is characterized by two operational states and three energy flows. The input energy 1 is influenced by two parameters $I_{1,1}$ and $I_{1,2}$, which do not relate to an operational state. In contrast to that, the influential parameter for energy flow 2, $I_{21,1}$ is only relevant for the first operational state. The output energy flow 3 represents an example with influential parameters for both the entire system ($I_{3,1},\ldots,,I_{3,r}$) and the different operational states ($I_{31,1},\ldots,,I_{31,p}$ and $I_{32,1},\ldots,,I_{32,q}$, respectively).

The application of the method requires identifying relevant parameters for each technical object system. Although the influence is formulated as an equation, the intensity does not need to be specified (*e.g.*, velocity is squarely included in the calculation of kinetic energy). Instead, the equations represent qualitative interrelationships in this context. However, general physical laws and equations might be helpful to determine relevant parameters.

In the following sections, this approach is presented in more detail with reference to relevant object systems in a factory. Therefore, the next sections are classified similarly as the hierarchical description of factory systems.[6] The purpose is to demonstrate which areas need to be considered for diverse object systems and how to generate an extended functional model. Based on these explanations, the energy efficiency influential parameters need to be specified for a concrete planning task as part of the method's application.

4.4.2 Influential Parameters on the Energy Efficiency of Buildings

Models for the energy consumption of buildings can be roughly classified into detailed approaches with complex functions including physical effects on the one hand, and simplified approaches on the other hand (Zhao & Magoulès, 2012, p. 3587). The former

---

[6] The hierarchical levels division and segment are not explained separately since these can be represented by the combination of several work centers.

category allows a precise calculation, whereas the latter one is intended for a rough prediction of building energy consumption.

As an example for a simplified model, LÜ ET AL. suggest a differential equation for the air temperature within a building (*i.e.*, the change in temperature over time) depending on the factors building envelope, solar load, ventilation load, load due to occupancy, lighting, and appliance loads (Lü et al., 2015, p. 263). Since even the application of simplified models requires extensive expert knowledge and time effort, a variety of software tools exists to support the modeling of building energy consumption (International Building Performance Simulation Association, IBPSA, 2016).

International norms and standards support the calculation of a building's energy consumption. An indicator-based approach is described in VDI guideline 3807 by using characteristic values for the consumption of end energy, water, and electricity of a building. The values are related to a reference area in order to enable a comparison between different buildings. Therefore, a building is characterized by the following five indicators: the specific thermal energy consumption value e_{VT} in kWh/(m$^2 \cdot$a), the specific heating energy consumption value e_{VS} in kWh/(m$^2 \cdot$a), the specific electricity consumption value e_{VS} in kWh/(m$^2 \cdot$a), specific water consumption value v_{VW} in l/(m$^2 \cdot$a), and the final energy consumption value E_{Vg} in kWh. Every value is adjusted according to the outside temperature. (VDI 3807, p. 10)

Verifying the energy consumption of buildings is important for legal considerations. The Energy Saving Ordinance (EnEV) requires building owners to comply with requirements in order to receive a building permit: As such, the yearly primary energy consumption shall not exceed the energy consumption of a reference building with the same dimensions. Furthermore, the heat transfer coefficients of the external surfaces are limited to defined values. Calculating these values follows standardized procedures (DIN 4701; DIN 4108; DIN V 18599, Part 1).

The primary energy consumption is composed of the energy consumption resulting from different energy carriers, each of which multiplied with a factor that represents the environmental effects of this energy carrier. This environmental assessment considers relevant pre-processes for an energy carrier, such as the generation, conversion, and transport. As an example, renewable energies (*e.g.*, solar energy) are accounted for with a factor of 1 (DIN V 18599, Part 1, p. 67). The general mix of electricity as produced in Germany is considered by a factor of 2.8 (DIN V 18599, Part 1, p. 67). The end energy consumption is composed of the usable energy and energy losses due to generating, transporting, and storing energy. The usable energy of a building consists of the energy consumption for heating, cooling, air-treatment (*e.g.*, ventilation, humidification), lighting, other media supply (such as compressed air), and hot water systems (DIN V 18599, Part

1, pp. 24 ff.). Important influencing factors on the energy efficiency for these subsystems are explained in the following paragraphs.

Heating and Cooling

The need for heating and cooling is influenced by heat input into and output from the building. The demand for heat $Q_{h,b}$ results from the difference between heat sinks Q_{sink} and heat sources Q_{source} (DIN V 18599, Part 1, p. 30):

$$Q_{h,b} = Q_{sink} - \eta \cdot Q_{source}, \tag{4.8}$$

whereby η characterizes a utilization factor that accounts for the alteration of heat sinks and sources over time. General external influential factors for the heating and cooling demand are climatic conditions (*e.g.*, outside temperature), use of the building (*e.g.*, room temperature), and the behavior of users (*e.g.*, frequency of window ventilation).

The heating or cooling demand is derived from the balance between heat sinks and sources. This includes transmission heat sinks, ventilation heat sinks, radiation heat sources, internal heat sources, and heat that is stored on normal usage days and released on days with reduced building usage.

Transmission heat transfer occurs between the building and its surrounding area and depends on the transmission heat transfer coefficient of the building material, the cross-sectional area of the respective structural element, and the difference between room temperature and outside temperature. Thermal bridges need to be considered as well. These occur in the cases where a structural element has a higher heat transfer than the surrounding elements. An example for this situation are building corners since a small inside area is opposed to a large outside area leading to a significantly higher heat transfer. Transmission heat losses, in general, need to be calculated for outside structural elements and for the heat transfer into soils. In the case of unheated or cooler adjacent rooms, a calculation is also required for inside elements.

Ventilation heat sinks or sources are caused by air exchange between room air and outside air or between zones with different temperatures in a building. Ventilation heat losses need to be considered for exchange with outside air (*e.g.*, open gates), window ventilation, mechanical ventilation equipment, or air exchange with other building zones.

Solar heat input originates from both transparent and opaque structural elements and is influenced by the area of the element, the solar intensity of this geographic region, and the geometric orientation of the element (*e.g.*, angle). Other influences may be caused by solar protection or shading. The heat input results from the balance between absorbed and emitted radiation. The absorbed radiation depends on material and color, while the emitted radiation additionally depends on the average difference between the surrounding temperature and the (assumed) temperature of the sky.

Internal heat sources or sinks may include electrical equipment (*e.g.*, machinery, lighting), persons, or material (*e.g.*, goods with high temperature). Lighting is usually completely effective as a heat source. In the case of material and goods, the emitted heat depends on mass, heat capacity, and difference in temperature.

The share of heat sources that is not used for heating purposes needs to be considered for the cooling need. In the case a building is not cooled, this share results in higher room temperatures or is compensated by increasing heat sinks such as additional window ventilation. If the building usage varies significantly, energy balances may be calculated for various use case scenarios. For instance, these may distinguish between week-days and weekends.

The demand for heat $Q_{h,b}$ resulting from the energy balance forms the input to calculate the heating energy consumption, which is influenced by generation, storage, distribution, and heat transfer into the room.

Heat generation is required in order to fulfill the heat energy demand in a building. It may be reduced by heat generated through air-conditioning or solar systems. Several technologies are available for heat generation (*e.g.*, heating boilers, district heating, combined heat and power unit). Heat storage may be required in the cases where the occurrence and demand for heat do not arise simultaneously (*e.g.*, solar heat generators). The use of heat storage units leads to storage losses. These are affected by storage temperature, surrounding temperature, and heat loss in stand-by mode, which is mainly influenced by the storage size.

The energy losses due to the distribution of heat in pipe systems mainly depend on the pipe length, the temperature of the heating medium, and the surrounding temperature. Supporting energy demand for the heat distribution is required for the circulation of the heat medium with pumps. The energy consumption for circulation depends on the efficiency and usage of the heating pump (*e.g.*, efficiency during periods of partial load).

The heat transfer into the room is characterized by a system efficiency. For example, warm air systems and radiation heating systems differ in terms of the vertical profile of room temperature.

Air Conditioning

The energy consumption for air conditioning is composed of two components: At first, energy is necessary to produce an air flow, for example by a fan. Secondly, the air needs to be conditioned depending on the requirements of the production process. This includes heating, cooling, humidifying, and/or drying air.

The production of an air flow consists of supplying fresh air and draining exhaust air. The electrical power that is necessary for this task depends on the volume flow, the pressure losses of the pipe network, and the efficiency of the air system that includes fan, pipe network, motor, and speed control.

Similar to the heating and cooling system, air conditioning consists of systems for generation, transfer, distribution, and storage. Losses in the distribution network may occur in the cases where the air ducts are located outside of the building, *i.e.*, with a lower surrounding temperature, or when leakages are present. However, the main influence of the energy consumption for air conditioning is the specification of requirements (*e.g.*, aspired air humidity).

Lighting

The energy consumption for lighting may vary between different zones in a building, for example due to the availability of daylight or different usage times. Daylight use depends on the window material, room depth, sun protection, anti-glare filters, pollution of transparent elements, and effects that hinder the daylight from entering the building (*e.g.*, trees in front of a window). The lighting electrical power is affected by the illumination intensity, the efficiency of the lamp, the efficiency of the illuminant in the lamp, and the lighting efficiency of the room.

A major influence on lighting energy consumption results from the desired illumination intensity. This should be adjusted to the requirements of the production process (*e.g.*, low illumination for transport paths, high illumination for quality control). Furthermore, the lighting should be generated efficiently, especially by selecting efficient illuminants.

Compressed Air Systems

Compressed air is an energy carrier that is widely applied in the manufacturing industry. The average share of energy consumption in an enterprise that is required to generate compressed air accounts for approximately 10 % (Bayerisches Landesamt für Umwelt, 2009, p. 18). A compressed air system consists of the areas generation, distribution, and usage. The energy that is required to generate compressed air depends on the efficiency of the compressor and the required air quality (*e.g.*, dry, oil-free). Moreover, the generation needs to be dimensioned properly according to the requirements of the production process. The excess heat of compressors may be used for heating purposes (*e.g.*, hot water generation).

In compressed air distribution, leakages are a major cause of energy losses. Furthermore, energy losses are influenced by the pipe diameter and the applied fittings. The usage of compressed air should be limited to applications, in which its advantages are exploited, *e.g.*, holding a position (Hülsmann et al., 2012, pp. 22 ff.).

Hot Water Systems

The energy demand for hot water mainly depends on the water volume and the tap water temperature. The energy consumption for heat generation depends on the applied technology (*e.g.*, solar systems, heat pumps, heating boilers, or distant heat). Additionally, energy losses occur due to the distribution and storage of hot water. Similar to the losses for heat distribution, the energy losses of hot water distribution are mainly influenced by the pipe length, temperature, and heat transition coefficient. Additionally, supporting energy is required for hot water circulation and loading the hot water storage.

Power Generation

Power generation may be integrated into a building, for example by means of a combined heat and power unit, wind power system, or photovoltaic system. If a combined heat and power unit is used, heat is produced besides the power generation. This reduces the heat demand for conventional heating systems.

Building Automation

Building automation might be used in order to adjust the energy use of a building to its actual usage. For example, the temperature can be reduced during the absence of persons in a room (*e.g.*, at night). Intelligent control may reduce the time, during which energy is consumed, or the power need of a system.

4.4.3 Influential Parameters on the Energy Efficiency of Work Centers

A work center is characterized by an individual, specific purpose towards the material or energy flow. First, general remarks on describing the energy efficiency of work centers are discussed. Afterwards, an extended functional model is generated exemplary for a cutting process.

General Influential Parameters of Work Centers

The energy consumption of a work center is characterized in general by the operational states off, stand-by, setup, error, block, wait, warm-up, work, and (optional) an energy-saving mode (Haag, 2013, pp. 75 f.). This variety of states is not necessarily considered for each work center, in case the power level is the same for several operational states. The energy consumption of production processes is commonly defined by the states base level, ready state, and operating (Balogun & Mativenga, 2013, p. 180). The base level represents the minimum power demand of a work center, often referred to as stand-by mode. It may occur, for example, during breaks or when a machine is set up. The ready state means that a work center is ready to perform its purpose (*e.g.*, waiting for a part). The operating level represents the power demand that is necessary to perform the purpose of the work center (*e.g.*, machining, transporting). For peripheral systems, the variety of operational states is typically lower (Haag, 2013, p. 77).

There might be several operating power levels depending on the utilization rate of the work center (Hopf & Müller, 2013b, p. 1669). For example, the power demand of a conveyor varies with the velocity and mass of transported goods. The energy consumption within any operational state is characterized by the power demand of this state and the time that the work center spends in this state (see Section 4.4.1).

Another possibility to describe the energy consumption of a work center is to differentiate it into its subsystems. As such, the model for the total energy consumption can be distinguished into work for the main process, work for auxiliary devices, and losses work (Müller et al., 2009, p. 71). The work for the main process includes the energy demand for the desired output of a system (*e.g.*, machining). The work for auxiliary devices covers energy demand for all other facilities, *e.g.*, control, lubrication system. The losses work includes the physical work due to conversion between different energy forms (*e.g.*, heat losses of a motor).

Example: Energy Consumption of Cutting Processes

Cutting processes refer to manufacturing processes where the cohesion of material is removed (DIN 8580, p. 4). The initial product contains the final product of a cutting process. The energy consumption of a machine tool may be divided into the energy for the cutting process (tool tip energy), and the energy for peripheral units (Salonitis & Ball, 2013, p. 636). Furthermore, the peripheral energy consumption can be distinguished into background energy, which depends on the machine tool, and load-dependent energy demand, which depends on the specifics of the process (*e.g.*, process parameters). Another aspect of the peripheral energy consumption is whether it is required to perform any operation (*e.g.*, tool movement) or converted to heat, *i.e.*, unproductive energy (Li & Kara, 2011, pp. 1642 f.). The corresponding operational states of a machine tool are base, idle, and cutting operation (Balogun & Mativenga, 2013, p. 182).

The total energy consumption E of a cutting process may be calculated as (Balogun & Mativenga, 2013, p. 182):

$$E = P_b \cdot t_b + (P_b + P_r) \cdot t_r + P_{air} \cdot t_{air} + (P_b + P_r + P_{cool} + k \cdot \dot{v}) \cdot t_c. \qquad (4.9)$$

The power demand levels comprise the basic power (P_b), ready state power (P_r), average power requirement for non-cutting approach (P_{air}), and coolant pumping power (P_{cool}). The respective time periods are basic time (t_b), ready state time (t_r), time for a non-cutting approach (t_{air}), and cutting time (t_c). The basic power represents the energy demand while the machine tool is switched on, whereas the ready state includes bringing the tool to the position where it is about to cut. The parameter k describes the specific cutting energy in kJ/cm³, which depends on the workpiece machinability and the cutting process parameters. The parameter $\dot{v}$ represents the rate of material processing in cm³/s, referred to as material removal rate.

The energy efficiency in cutting may be determined in terms of the specific cutting energy e, *i.e.*, the energy that is necessary to remove a certain volume of material (Pfefferkorn et al., 2009, p. 358):

$$e = \frac{P_{\text{tool}}}{\dot{v}} = \frac{F_C \cdot V}{\dot{v}}, \tag{4.10}$$

with P_{tool} being the mechanical cutting power as composed of the main cutting force F_C and the cutting velocity V. The specific cutting energy depends on workpiece material, cutting conditions, tool material and geometry, and temperature. Together, an extended functional model is generated based on the operational states and the energy efficiency influential parameters (see Section 4.4.1). The extended functional model for cutting processes is depicted in Figure 24.

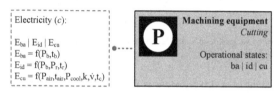

Figure 24: Extended functional model of a cutting process

As an example for the energy consumption of a cutting process, Figure 25 shows the power demand of a 3-axis milling machine.[7] At first, the machine is in stand-by mode and has a basic power level of 350 W. After 10 seconds of stand-by, the spindle starts rotating and moving towards the position which is about to cut. When starting this step of a process, a peak power demand of 2,600 W is reached. The actual cutting process is performed with an average power demand of 545 W and takes 94 seconds to complete a groove of 152 mm length. Afterwards, the tool moves back into its initial position. The stopping movement of the spindle leads to a short power demand peak. At the end of the process, the machine is in stand-by mode again. The total energy consumption of this cycle is 66 kWs with a material removal rate of 23.6 mm³/s.

4.4.4 Influential Parameters on the Energy Efficiency of Components

The energy consumption of a component as the smallest unit in the hierarchical system structure forms the basis for the energy consumption of work centers. The distinction between a work center and its component is made based on the individual purpose

---

[7] The respective machine tool is located in the Laser-Assisted Multi-Scale Manufacturing Lab at the University of Wisconsin-Madison. The measurements were taken in July 2014.

towards the product flow (see Section 4.3.2). Moreover, the separate description of a component is indicated when an energy flow of the work center might be assigned to a single component. In this context, relevant components comprise drive systems and pump systems (Brecher et al., 2014, pp. 495 ff.).[8]

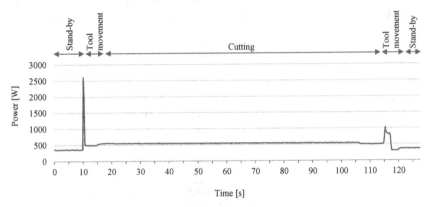

Figure 25: Exemplary power profile of a milling machine reflecting the energy consumption structure of cutting processes

The main component of a drive system is the motor that converts electrical energy into mechanical energy. The gearbox transfers the mechanical energy from the motor depending on the requirements of the work machine (*e.g.*, in terms of rotational speed and torque). The electrical power demand of a motor is composed of active power and losses (*e.g.*, due to friction). The ratio between provided mechanical power and supplied electrical power is the motor efficiency (Deutsche Energie-Agentur GmbH, dena, 2010b, p. 3).

The efficiency of a motor depends – besides its internal components – on the motor load L_{motor}, which is expressed as ratio between shaft power and nominal motor power (Kuhrke, 2011, p. 69):

$$L_{\text{motor}} = \frac{P_{\text{shaft}}}{P_N}. \tag{4.11}$$

Using this value, the motor efficiency can approximately be calculated as (Kuhrke, 2011, p. 71):

$$\eta_{\text{motor}} = \frac{L_{\text{motor}}}{L_{\text{motor}} + k_0 + k_1 \cdot L_{\text{motor}}^2}, \tag{4.12}$$

---

[8] It is assumed that 70 % of the electricity consumption in industry are caused by drive systems (Bayerisches Landesamt für Umwelt, 2009, p. 14). A high share relates to the usage of pump systems that accounts for approximately 25 % of the energy consumption in industry (Bauernhansl et al., 2014, p. 84).

whereby the coefficients k_0 and k_1 need to be estimated for each motor taking the values of 50 %, 75 %, or 100 % load.[9]

An example for the energy consumption of motors is given by the drives of a conveying system.[10] The exemplary considered motor has a nominal power of 370 W and an efficiency of 73 % at 75 % and 100 % load. It is equipped with a bevel gear that has an efficiency of 96 %. The curve of the efficiency as calculated by Equation (4.12) is depicted in Figure 26. The actual load depends on the mass of the transported goods.

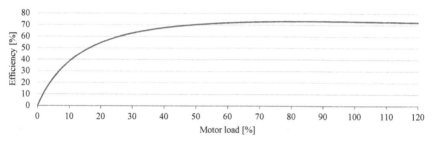

Figure 26: Exemplary efficiency of an asynchronous motor depending on the motor load

Pumps are used to convert mechanical energy into hydraulic energy and are used in various applications, such as transportation of fluids in heating systems or process facilities (Brecher et al., 2014, p. 516). Reducing their energy consumption is important since the energy costs have a high share of between 45 and 85 % of the life cycle costs (Blesl & Kessler, 2013, p. 54).

The hydraulic power of a pump depends on the pressure difference and the volume flow (DIN V 18599, Part 8, p. 33). The efficiency of a pump is calculated by (Watter, 2015, p. 99):

$$\eta_{pump} = \eta_{vol} \cdot \eta_{hyd} \cdot \eta_{mech}, \tag{4.13}$$

whereby η_{vol} accounts for volumetric losses (*e.g.*, leakage), η_{hyd} for hydraulic losses (*e.g.*, flow losses, fluid friction), and η_{mech} for mechanical losses (*e.g.*, bearing friction).

4.5 Description of the Project Characteristics

The goal of modeling the project characteristics is to capture the soft aspects of the energy efficiency improvement task since they imply important conditions for identifying energy efficiency measures. A project is characterized by its relative novelty. However, projects

---

[9] Motor manufacturers are required to specify the efficiency at 75 % and 50 % of nominal load according to the Regulation No. 640/2009 of the European Commission (European Commission, 2009a, p. 31).

[10] The conveying system is part of the Experimental and Digital Factory at Chemnitz University of Technology. Energy measurements were conducted as part of a research project in January 2014.

may have similarities and can be classified into categories. Table 7 shows ten dimensions for describing project types.

Table 7: Project characteristics (Gessler, 2014, pp. 43 ff.)

Dimension	Project type
Ordering party	External, internal
Business value	Strategic, tactical
Project content	Investment, research and development, organizational development
Relative novelty	Innovation, specialist, routine
Complexity	Standard, acceptance, capability, pioneer
Project organization	Influence, matrix, autonomous
Project control	Technocratic, agile
Geography	National, international
Project size	Small, medium, large
Project role	Client, contractor

The project characteristics that need to be defined focus either on the factory planning project or the energy efficiency project.[11] Factory planning projects are characterized by planning phases, namely setting of objectives, establishment of the project basis, concept planning, detailed planning, preparation for realization, monitoring of realization, ramp-up support, and planning steps, namely determination of functions, dimensioning, structuring, and design (see Section 3.4.3). The possibilities to influence energy consumption vary depending on the planning activity. For example, organizing the location of equipment within a factory in order to reduce the length of media pipes (*e.g.*, for cooling water or compressed air) is possible within the planning step structuring. Considering the planning phases and steps as part of the method is important, since the influence on the energy consumption of a factory is higher in early planning stages (Engelmann, Strauch & Müller, 2008, p. 61). This is due to the fact that planning tasks in the early stages of the factory life cycle set the basis for the energy consumption during operation.

Moreover, the factory life cycle needs to be addressed, because the influences on energy efficiency vary with the life cycle phases. In general, a life cycle represents the development of an object from its origin to its degradation. Hence, the factory life cycle contains the period of time from its development through the dismantling. The factory life cycle comprises the phases development (planning), construction (realization), ramp-up (installation), operation (utilization), and dismantling (recovery) (Schenk, Wirth & Müller, 2014, p. 150).

---

[11] In this context, the energy efficiency project is interpreted as part of the factory planning project that aims at increasing the energy efficiency.

While distinguishing the relevant object areas, another model for the factory life cycle consists of investment, site and network, building, logistics and layout, processes, equipment, ramp-up, operation, maintenance, and dismantling (Landherr et al., 2013, p. 169).

The factory life cycle may be described as the interaction between the life cycles of the area, building, process, and product (Schenk, Wirth & Müller, 2014, pp. 147 f.). The process life cycle is designed according to the product life cycle and influences the life cycle of production equipment.

The life cycle of the area is the longest period of time considered in factory planning and is most important in location planning. Land management considers the usage of areas as permanent cycle of planning, usage, abandonment, fallow, and re-use (Deutsches Institut für Urbanistik, 2006). However, the latter tasks are usually in the responsibility of municipal authorities. From the perspective of factory planning, the main difference is between the planning phase ("greenfield projects") and the usage phase ("brownfield projects").

A building's life cycle consists of four phases: production, construction, operation, and demolition. The production phase includes the procurement and transport of raw building material. During the construction phase, the building is raised. The operation phase includes maintenance, repair, replacement, and modernization. These activities are understood as continuous processes rather than being conducted successively. However, the effort and extent of change varies between these processes which may affect the frequency of the activities. Maintenance refers to processes, in which components of the building are restored to their initial state without a prior damage (*e.g.*, changing a filter in the air conditioning system). Repair means to restore the initial state after a damage (*e.g.*, repairing a broken window glass), whereas replacement means exchanging a component (*e.g.*, replacing a window). The modernization is the most comprehensive activity during the operation phase of a building (*e.g.*, adjustment of the heating system in preparation of a new concept for room usage). (DIN EN 15978, p. 21)

The life cycle of production equipment consists of the phases concept, development, manufacturing, installation, operation (including maintenance and modernization), and disposal. During concept definition, requirements for the equipment are defined and analyzed. The purpose of the development is to create the design, starting from a draft of principles and finishing with a detailed design drawing. Afterwards, equipment is produced (manufacturing), set up (installation), and used in a factory (operation). (DIN EN 16646, p. 20)

The VDI guideline 2884 considers the life cycle of production equipment from the perspective of the manufacturer and from the perspective of the operator (VDI 2884, p. 5): From the view of the manufacturer, the life cycle contains initiation, planning, development, and realization of equipment. The perspective of the operator differentiates between prior to operation, operation, and after operation. Previous to the operation of production equipment, activities such as the projecting, procurement, and installation are performed, whereas projecting means to describe the requirements in a specification sheet. Based on this, offers are invited as part of the procurement. The installation includes setting up the equipment in the factory. During the utilization phase, the equipment is operated and maintained in order to perpetuate its performance. Afterwards, the equipment is discarded and disposed (Pohl, 2013, p. 34).

The life cycle of a product is divided into the phases introduction, growth, maturity, saturation, and decline from an economic perspective; the phases differ in terms of product sales (Herrmann, 2009, p. 70).

From a technical perspective, the product life cycle begins with the product development and ends with the product disposal, containing the phases product planning, design, work preparation, production planning, manufacturing, sales, service, and recycling (Schuh & Stich, 2012, p. 340). The interrelationships between the product life cycle and the energy-efficient design of factory systems result from the fact that requirements from the product specify the production processes. Figure 27 summarizes the life cycles of area, building, and equipment including their interrelationships.

The characteristics of the energy efficiency project determine the opportunities for energy efficiency measures. The project scope or size may be expressed in terms of budget and/or time. KESSLER and WINKELHOFER classify projects depending on their budget into projects with financial resources less than 100,000 Euro, less than 1 billion Euro, less than 10 billion Euro, less than 100 billion Euro, and more than 100 billion Euro (Keßler & Winkelhofer, 2004, p. 34). It should be noted that this classification is very general and may be adapted depending on the type of project. The project time refers to the time that is planned for the implementation of energy efficiency measures. A possible classification distinguishes between less than one month, one to six months, 6 months to one year, one to two years, and more than two years (Keßler & Winkelhofer, 2004, p. 34). With relation to the industrial branch, the difference between mid-term and long-term projects is marked by the life cycle of the production equipment (Aggteleky, 1987, p. 29).

4.6 Description of the Actor

The actor is an important aspect for describing the project task in the context of socio-technical factory systems (see Section 3.4.2). The influence of a user varies depending on his or her role in the factory. For example, an operator of a machine has different opportunities than the person responsible for planning or programming this machine. Considering the user's role therefore helps to tie the identified measures to the personal scope of activities. Factory planning projects are characterized as being highly interdisciplinary, which increases the importance to describe the different actors in factory planning and management.

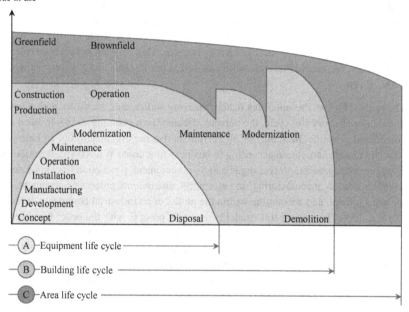

Figure 27: Interrelationships between equipment, building, and area life cycle (adapted from Schenk, Wirth & Müller, 2014, p. 148)

In general, role models are used to represent the activities and relationships of different actors as part of the organizational structure (Esswein, 1993, p. 555). In the context of project management, a role model summarizes all stakeholders within that project (Broy & Kuhrmann, 2013, p. 42). Therefore, it allows to reflect the project organization and to provide the interface to the primary organization of an enterprise. An example for a role model is the organization chart of a company that displays the organizational structure (Scheer, 2002, pp. 17 f.). The person that holds a role is called a role owner. A person may hold several roles within an organization and one role may be held by different persons.

A role model may be formulated with regard to different perspectives (Eberhard, 2009, p. 86): A *competence-related role model* represents the qualifications that an employee requires to fulfill the tasks of his or her role. This information may be used to describe requirements for an applicant. The *task-related role model* characterizes a role by the entirety of tasks, which may be used to describe the responsibilities of a job position.

The *permission-related role model* describes the permissions of an employee, which is used to regulate the access to resources (*e.g.*, documents in an information system). An *organization-related role model* reflects the position of a role in the organizational structure (*e.g.*, regarding rules for substitutes). In a *behavioral role model*, a role is understood as the expectations from other stakeholders towards this position. Hence, this perspective focuses on the social interactions of role owners. Role models are usually graphically represented by means of a role diagram. This diagram represents a role with its activities, required skills, and abilities as well as permissions and obligations (Schütze, 2009, p. 100).

The method focuses on supporting factory planning participants. However, with reference to the entire factory life cycle, the persons affected from planning decisions need to be considered as well. These may be all persons in the line organization of an enterprise, which is usually structured according to business functions. WIENDAHL describes the functions management, process organization, procurement, production (including design, work preparation, manufacturing, and assembly), distribution, process management, quality management, and accounting within the model of an industrial company (Wiendahl, 2014, p. 25). The core of this model is the main process with the order flow through the factory (*i.e.*, from procurement through production to distribution). The process organization describes methods and tools for realizing these functions while the process management has an operative task to manage the activities. The functions may be detailed further on, for example by adding the marketing department to the area of sales or by maintenance engineers as part of the production department. Moreover, indirect functions such as human resources, information technology, and legal affairs may be considered.

Besides the differentiation of functions, the organizational structure describes hierarchical relationships, *i.e.*, permissions to issue instructions. For example, the Toyota Production System defines the following roles in manufacturing and assembly areas: The *team members* fulfill standardized tasks in the production system while continuously searching for improvement opportunities. A *team leader* is leading the production teams (*e.g.*, for the section of a line) and substitutes the team members. The *group leaders* are on a supervising level and lead a small number of groups. The next hierarchical level is the *assistant manager* who is responsible for a production area. Finally, the *department manager* is the head of the production plant. (Liker, 2007, p. 273)

As part of a production system, operative team members may differ regarding their technical responsibilities and task complexity. For example, the labor union distinguishes between the following roles in mechanical manufacturing: The *machine tender* puts the parts into a set machine and removes them after an automatic process. A more complex function is fulfilled by the *machine operator*, who additionally provides necessary material and checks the product quality. The *machine setter* needs to set up the machine including minor adaptations of the machining program and change of tools depending on the product requirements. The *machine supervisor* is responsible for the entire machine including work preparation (*e.g.*, availability of tools and material), optimizing machining procedures, supervising machining, and measurement processes. (IG Metall Baden-Württemberg, 2003, pp. 190 ff.)

Factory planning projects are usually carried out in interdisciplinary teams (see Section 3.4.3). Hence, representatives of different primary divisions may become members of the planning project team: The *business manager* represents the corporate management of the enterprise and has the role of the ordering party for the project. Thus, he or she defines the goals for the project, selects the project manager and provides the required project resources. The *product designers* define the product specifications (*e.g.*, material, geometry), which are important input information for the factory planning process. The *sales representatives* contribute to the definition of the production and performance program (see factory planning steps in Section 3.4.3). *Procurement employees* may have a similar role. Furthermore, this actor influences the design of the factory by inviting offers (*e.g.*, for production equipment). The *human resources department* is involved, for example, when defining requirements for the factory staff. The *marketing representatives* analyze the market requirements, which supports the definition of the production program. *Accountants* are responsible for the budget of the project and fulfill financial project tasks, such as conducting cost analyses and calculating investments. (Günther, 2005, p. 13; Schenk, Wirth & Müller, 2010, p. 18; Grundig, 2015, p. 20)

Besides, a factory planning project may contain the following roles (Felix, 1998, pp. 789 ff.; Grundig, 2015, p. 20; Schulte, 2009, p. 88; Günther, 2005, p. 13; Schenk, Wirth & Müller, 2010, p. 18; Schenk, Wirth & Müller, 2014, p. 25):

- civil engineer,

- architect,

- production engineer,

- logistics planner,

- information engineer,

- ergonomic engineer,

- specialist engineer,

- planning method specialist, and

- project manager.

The *civil engineer* plans the technical aspects of the building and the production site. This includes the supporting structure and construction physics as well as special objectives, such as sound insulation and fire protection. The *architect* handles the design aspects of the building, such as the interior construction. Additionally, the architect drafts the building concept according to legal regulations (*e.g.*, waste water legislation).

The *production engineer* is involved in planning the manufacturing and assembly area. This includes projecting machinery and equipment as well as defining the corresponding processes. The *logistics planner* performs tasks with regard to transport, handling, storage, and order picking systems. The *information engineer* plans and operates the information and communication systems, such as control and automation systems. The *ergonomic engineer* supports the work preparation through applying working time systems and plans workplaces with regard to ergonomic criteria.

Specialist engineers may be integrated for more detailed topics (*e.g.*, consultant for environmental protection). A *specialist for planning methods* contributes detailed knowledge on methods and tools to support the planning process (*e.g.*, simulation, software for managing the planning process). Finally, a factory planning project is managed by a *project manager*. This role is responsible for monitoring and controlling the factory planning project towards the achievement of the project goals (budget, deadlines, and quality). Characteristic project management tasks include the project organization as well as planning of procedures, deadlines, resources, and costs.

Since the specific actors may differ for each use case, the following explanations provide a general description structure for the domain actor. The purpose is to describe the possibilities of a role to influence the energy consumption of an object system. Therefore, a task-related view is pursued.

A classification of the relation between a role and an activity is given by the so-called RASCI structure (Wysocki, 2011, pp. 66 f.):

- responsible (R): the role is realizing a task,

- accountable (A): the role is liable for the task's realization,

- supportive (S): the role supports the realization of a task,

- consulted (C): the role is consulted before a decision, and

- informed (I): the role needs to be notified after a decision has been made.

Taking this as a basis, the responsibility of each role towards an object system is described. Furthermore, the influence on the energy efficiency of relevant object systems needs to be described as part of the model. Each role is described by the following characteristics: First of all, the roles are assigned to relevant object systems, *i.e.*, the involvement of a role towards an object system. It should be noted that the description of relevant actors needs to integrate both factory planning participants and persons affected from planning decisions. Secondly, tasks and activities towards an object system are explained with a short description and may be classified following the RASCI structure in case there are tasks with several roles involved.

Based on this information, the effect on the energy efficiency influential parameters is described. This means that the influence of each role is assessed on the relevant object system and the corresponding parameters. The influence of an actor on the energy efficiency may change during the life cycle of an object. Hence, the influence is represented by a three-dimensional description including actors, life cycle phases, and energy efficiency influential parameters (Table 8).

Table 8: Description structure for the actors' influences on energy efficiency

Actor	Tasks	Planning phases			
		P_1	P_2	P_3	...
A_1	...	$I_{1,1,1},..., I_{1,1,p_1}$	$I_{1,2,1},..., I_{1,2,p_2}$	$I_{1,3,1},..., I_{1,3,p_3}$	...
A_2	...	$I_{2,1,1},..., I_{2,1,p_1}$	$I_{2,2,1},..., I_{2,2,p_2}$	$I_{2,3,1},..., I_{2,3,p_3}$	...
...	...	...	...	...	...

It should be noted that the considered life cycle phases may vary depending on the object system. For example, the influence of the role building planning varies with the phases of the building life cycle, whereas the influence of a logistics planner may be better described with reference to the factory planning phases.

4.7 Description of the Energy Efficiency Measures

The energy efficiency measures are part of the knowledge that is to be assigned to the factory planning task. After specifying the definition of measures, a literature review is conducted in order to identify appropriate classifying criteria. Then, the structuring criteria for the method are deduced and their application to represent the measures in the knowledge base is explained.

Definition of Energy Efficiency Measures

An energy efficiency measure (EEM) describes a solution approach that is to be identified by the methodical approach. In the following, an EEM is understood as a measure that may include technological, planning, behavioral, economic, and organizational changes, and aims at improving the energy efficiency (adapted from the definition of an energy performance improvement action, see DIN EN ISO 13273, p. 11). As introduced in Section 2.1, energy efficiency is defined as the ratio between an output and an energy input. Within this thesis, the goal is to reduce energy consumption while maintaining the useful output. This means that an EEM shall not generally include a decrease in process output, although it may have this effect in a specific use case.

Measures are classified depending on whether they directly influence the energy consumption (DIN EN 16212, p. 8). An example for a direct action is the usage of energy-efficient motors in a machine (*i.e.*, directly influencing the efficiency factor); an exemplary supporting measure is the use of an energy monitoring system that regularly checks a machine's energy consumption, which may be used as basis to determine further measures. Since the assignment task of the method is based on energy efficiency influential parameters, the following explanations are limited to direct energy efficiency measures.

Literature Review on Structuring Energy Efficiency Measures

The purpose is to provide classifying characteristics for the EEMs which are used for the assignment to the factory planning project. Therefore, a literature review has been performed in order to identify classifications of energy efficiency measures.[12]

The most general criterion is the type of an EEM. A rough classification differentiates measures that need an investment and those that depend on the usage of equipment (Bürger, 2010, p. 49). NEUGEBAUER ET AL. divide measures into product-oriented, technical, and organizational (Neugebauer et al., 2010, p. 798). Additionally, social measures may be considered that represent the influence of the staff, *e.g.*, through motivational aspects (Dombrowski, Kynast & Aurich, 2012, p. 597). Furthermore, the mechanism of a measure may be used for classification, such as prevention, reuse, recycling, or recovery (Fischer, 2013, p. 107).

Several criteria may be used to describe the applicability of EEMs. This includes the energy type, such as electricity or heat (Trianni, Cagno & Donatis, 2014, p. 210), the specificity of a measure, *i.e.*, whether it is specific for a sector or may be applied industry-wide (Fleiter, Hirzel & Worrell, 2012, p. 508), the addressed life-cycle stage (Fischer, 2013, p. 105), and the distance to the core process (Fleiter, Hirzel & Worrell, 2012, p. 507). The specificity of an EEM is related to whether a measure may be transferred to other machines or areas (Böhner, Kübler & Steinhilper, 2013).

Another important dimension of an EEM considers its implementation. The implementation complexity of a measure may be described by the changes that are necessary for its realization. Similar approaches are used across literature to describe this criterion. BÖHNER ET AL. develop the stairs of innovation, differentiating between the characteristics substitution, upgrading, reengineering, product design, and technology management (Böhner, Kübler & Steinhilper, 2013). Whereas these only focus on technical measures, another opportunity is to address the novelty in general. FLEITER ET AL. distinguish

---

[12]The following presentation of the literature review has been published similarly in Krones & Müller, 2014b.

organizational measures, technology add-ons, technology replacements, and technology substitutions (Fleiter, Hirzel & Worrell, 2012, p. 507). In context to this criterion is the implementation time, *i.e.*, the time that is needed for planning and implementing a measure (Fischer, 2013, p. 106).

The connection between EEMs and the persons that can realize them may be addressed by the organizational structure or the required qualification. FISCHER provides a classification into management, corporate culture, human resources, research and development, product design, marketing, controlling, procurement, manufacturing, maintenance and cleaning, storage and logistics, and packaging (Fischer, 2013, p. 105). FLEITER ET AL. analyze the knowledge requirements for energy efficiency measures in order to describe their adoption rate (Fleiter, Hirzel & Worrell, 2012, p. 508): They differentiate into technology expert, engineering personnel, and maintenance personnel.

The assessment of measures is usually performed by evaluating the relation between costs and benefits. The costs for realizing a measure are influenced by the initial investment (Böhner, Kübler & Steinhilper, 2013), transaction costs (Fleiter, Hirzel & Worrell, 2012, p. 508), and the necessity of periodical check-ups, such as regular cleaning (Trianni, Cagno & Donatis, 2014, p. 212). These effects may need to be considered during the entire measure duration (Fischer, 2013, p. 106). Measures range from short-term (< 5 years) through medium-term (5-20 years) to long-term (> 20 years) duration (Fleiter, Hirzel & Worrell, 2012, p. 508).

The direct benefit of an EEM may be quantified as energy savings. With regard to ecological objectives, reducing emissions and/or waste may be another benefit (Trianni, Cagno & Donatis, 2014, p. 210). Besides, other business objectives may be influenced, such as increasing productivity or decreasing maintenance need (Fleiter, Hirzel & Worrell, 2012, p. 506). The cost-benefit ratio is economically assessed, usually by means of the pay-back time (Trianni, Cagno & Donatis, 2014, p. 210) or the internal rate of return (Fleiter, Hirzel & Worrell, 2012, p. 506).

As a result of the literature review, the characteristics to classify EEMs may be assorted in three groups (Table 9): The first one is used to identify proper energy efficiency measures depending on technical and organizational aspects. This includes the criteria energy carrier, application area (*e.g.*, procurement, logistics), life cycle phase (*e.g.*, planning, operation), and distance to the core process. The second group of criteria is used for the assessment of costs and benefits. It comprises investment, pay-back time, energy savings, emission savings, waste savings, influence on other objectives (*e.g.*, productivity), check-up frequency, and implementation time. The third group of criteria serves the purpose to prepare the measure implementation. It contains the implementation complexity (*e.g.*, required competencies), extent of changes (*e.g.*, new processes within the organization), transferability, and corporate involvement.

Table 9: Result of literature review on classifying criteria for energy efficiency measures

Criteria group	Criterion	Measurement type
Identification of EEM	Energy form or carrier	Nominal
	Organizational unit	Nominal
	Application area	Nominal
	Life cycle stage	Nominal
	Distance to core process	Ordinal
Assessment of EEM	Investment	Metric, ordinal, and qualitative
	Pay-back time	Metric, ordinal, and qualitative
	Internal rate of return	Metric and qualitative
	Transaction costs	Metric and qualitative
	Energy savings	Metric, ordinal, and qualitative
	Influence on other objectives	Qualitative
	Check-up frequency	Metric and qualitative
	Implementation time	Metric, ordinal, and qualitative
Support for realization of EEM	Realization complexity	Qualitative
	Extent of changes	Qualitative
	Transferability	Qualitative
	Effect on employees	Qualitative

Structuring Criteria for the Methodical Approach

In the following paragraphs, the classifying criteria that are used for the methodical approach are deduced from the previous findings. The method supports the identification of EEMs, their assessment based on general information and the preparation of their implementation (see Section 4.2). The first of these steps is supported by an algorithm that identifies suitable EEMs based on the requirements of a planning task (see Section 4.9). It should be noted that the algorithm can only process quantifiable information. As a prerequisite, the corresponding classifying characteristics of EEMs need to be formulated according to the description of the planning task (see Sections 4.3, 4.4, and 4.5).

The information to prepare a measure's implementation is part of the measure implementation support (see Section 4.8). The assessment is supported by both means: On the one hand, each measure should be provided with a quantitative information on the corresponding assessment criteria. On the other hand, there might be additional information which is only available in qualitative form. This will be integrated into the measure implementation support. An example is the quantification of energy consumption savings: The quantitative effect of a measure might not be available. However, qualitative information (*e.g.*, on which external factors the energy savings depend) might be provided for the user.

As a result, the first two groups, *i.e.*, identification and assessment of EEM, need to be implemented as classifying criteria in the description of the domain EEM.

The *industrial sector* defines the sector, in which a measure may be applied. The characteristics follow the classification as described in Section 4.3.1. Measures may be limited to specific industrial sectors or may be applied in general.

The *manufacturing type* distinguishes the general type of the production system, *i.e.*, depending on the product variety (see Section 4.3.1). The *manufacturing form* reflects the structural organization of production (see Section 4.3.1). For example, measures that address the stand-by power of machinery may have a higher effect in job shop production systems rather than in flow shop production systems since the latter ones have an optimized cycle time.

The *manufacturing principle* describes the spatial organization of production (see Section 4.3.1). For example, the workshop principle allows to group machinery depending on their media requirements such as compressed air.

The hierarchical model of factory systems is addressed since EEM may be realized on or may affect various *hierarchical levels*. This criterion follows the characteristics factory, building, division, segment, work center, and component as defined in Section 4.3.2. It should be noted that both realization and effect aspects need to be considered when evaluating an EEM. They may be the same for many EEMs. For example, the EEM "Reduce the velocity of logistics equipment" is realized on the level of a work center and also affects the energy consumption of the work center. On the contrary, the EEM "Reduce excess heat of machinery" is relevant for the hierarchical level of a work center (machine) and a segment (cooling system). This means that the measure needs to be considered when planning a machine but affects the energy consumption of the cooling system.

The *system function* considers the functional model of the factory system with regard to the material and energy flow (see Section 4.3.3). The *energy form* specifies the object of the energy flow.

The definition of the relevant *energy efficiency influential parameters* as described in Section 4.4 is an important characteristic to describe an EEM. This criterion is, among others, used to link an EEM to the relevant actor. It should be noted that the actor itself is not used as criterion to describe EEM since the model of actors may be specific for each enterprise or planning case.

The final criterion to support the identification of EEM refers to the *life cycle stage* of the area, building, equipment, and product or the factory planning phase (see Section 4.5). A summary of the identifying criteria is given in Table 10.

Table 10: Identification criteria to describe energy efficiency measures

Criterion	Characteristics
Industrial sector	Classification of industrial sectors
Manufacturing type	Individual production, series production, mass production
Manufacturing form	Single spot production, job shop production, group production, flow shop production
Manufacturing principle	Linear principle, parallel principle, nest principle, asterisk principle, workshop principle
Hierarchical level	Factory, building, division, segment, work center, component
System function (material flow)	Manufacturing systems, logistics systems, assembly and handling systems, building structure, property site, outdoor facilities, process technology, building services
System function (energy flow)	Provision, generation, conversion, storage, transport, consumptive usage, non-consumptive usage, recovery, emission
Energy form	Electricity, heat, excess heat, cold
Influential parameter	Selected from the energy efficiency influential parameters
Life-cycle stage	Area life cycle, building life cycle, equipment life cycle, product life cycle, factory planning phases and steps

For supporting the assessment of an EEM, the following criteria are used:

- energy savings,

- investment,

- pay-back time, and

- implementation time.

The quantification of the investment is a complex task since this is usually an absolute figure in terms of a currency value. However, whether an investment is considered "expensive" or "inexpensive" depends on a variety of factors. Therefore, the quantification is expressed with regard to the total investment budget of an enterprise (Fleiter, Hirzel & Worrell, 2012, p. 506). As a rough orientation, low investments relate to organizational measures, medium investments are limited to specific systems and high investments refer to the entire factory system. Table 11 depicts the ordinal description for the assessment criteria.

Matrix-Based Representation of Energy Efficiency Measures

Having specified the classifying criteria, their fulfillment needs to be assessed for the EEMs in form of a DMM (see Section 3.3.2) in order to prepare the assignment to the planning task. These DMMs form the knowledge base of the EEM. It should be noted that the relation between an EEM and the domain actor is not modeled directly since the knowledge base of EEM is independent from the enterprise-specific model of actors. Instead, the relation between an EEM and the energy efficiency influential parameters is represented. In the following paragraphs, the approach to determine the fulfillment values for the DMMs is explained.

Table 11: Assessment criteria to describe energy efficiency measures (adapted from Böhner, Kübler & Steinhilper, 2013; Fischer, 2013, p. 107; Fleiter, Hirzel & Worrell, 2012, pp. 506 f.)

Criterion	Characteristics		
	low	medium	high
Energy savings	< 10%	10 to 25 %	> 25 %
Investment	< 0.5 %	0.5 to 10 %	> 10 %
Pay-back time	< 2 years	2 to 4 years	> 4 years
Implementation time	< 6 months	6 months to 1 year	> 1 year

The elements of a DMM represent the intensity of the interdependence between the concepts that are described by the rows and columns of the matrix. These elements are quantified by the values 3 – great importance, 2 – medium importance, and 1 – low importance (see Section 3.3.2). In the case of no interdependence between the concepts, the value of this matrix element is defined as 0. This concept is adapted for the case that is considered here in order to reflect the possible relations between an EEM and its classifying criteria.

The relation between an EEM and any criterion may be characterized by one of the following:

- clearly defined,

- not decisive,

- decisive,

- not applied, or

- required.

In the first case, an unambiguous assignment of the EEM to one characteristic is possible. For example, the EEM "Cover process baths in order to reduce heat losses" clearly belongs to the product function of coating since the use of process baths is limited to this group of manufacturing processes. In this case, the value of the applying characteristic is assigned a 3 and the other characteristics a 0. In the second case, a criterion and its characteristics are not relevant to describe an EEM. As such, the EEM "Use frequency converters" does not depend on the manufacturing type, *i.e.*, may be applied in individual, series, or mass production. In this case, all characteristics are given the assessment 1.

The third case represents the situation, in which a criterion is crucial to characterize an EEM but a clear selection of one characteristic is not possible. As an example, the EEM "Insulate heating pipes" is relevant both for the functions building services and process technology. It should be noted that the importance of the characteristics may differ. For example, the EEM "Use energy-efficient motors" is more relevant in the case of planning a new system but may be appropriate for modifying an existing system as well. In this case, a distinction between more and less applying characteristics is done by assigning different assessment values (0, 1, 2, or 3).

The relation "not applied" appears when an EEM does not fit to any characteristic of a criterion. For example, measures that relate to the operation of a factory are not applicable in any of the factory planning phases. In this case, all values of the respective criterion are assessed with 0.

The last case describes the situation, in which a measure's application requires any criterion to be fulfilled. For example, the EEM "Reduce running time of pumps" is supposed to decrease the electricity consumption but may only be applied to systems that require fluids. In this case, a pre-evaluation is required to secure that the use case fulfills the minimum characteristics. Therefore, the relation between the EEM and the respective characteristic is given the value 4 and, hence, may be specifically considered by the matching algorithm.

The approach for this assessment is described exemplary for the EEM "Place windows in order to maximize daylight". This measure addresses the position and geometry of windows in the early planning phase of a building. The purpose is to use daylight in order to save energy consumption for artificial lighting. In the following, an example for each assessment category is given. The case "clearly defined" applies to the criterion energy form, since the measure is directly related to electricity as only energy form for lighting. The manufacturing form is a "not decisive" criterion since the measure may be applicable in single spot production, job shop production, group production, and flow shop production without any difference in importance.

The case "decisive" is true for the criterion hierarchical level. In general, the EEM refers to the entire building and the interrelationships of building structure and building services. However, it may also be applied only on the level of a segment, e.g., for the area with manual inspection tasks in an otherwise automated production area. Furthermore, it may be considered as a measure for the division building planning. Hence, the relations towards the characteristics building, division, and segment are assessed with a positive value. The system function is a "required" criterion. Although the measure influences the energy consumption of lighting, which belongs to the function of building services, it may be realized during building planning. Hence, it should only be suggested for a use case that contains planning the building structure. Therefore, the relation between the EEM and the characteristic building structure is evaluated by four, whereas the characteristic building services is assigned with a value of three. Table 12 summarizes the resulting assessment of the EEM for the selected criteria. It should be noted that not all characteristics are mentioned. This assessment represents a vector as part of the DMM between an EEM and its classifying criteria.

Table 12: Exemplary definition of classifying criteria for energy efficiency measure "Place windows in order to maximize daylight"

Criterion	Characteristics	Value for EEM
Manufacturing form	Single spot production	1
	Job shop production	1
	Group production	1
	Flow shop production	1
Hierarchical level	Factory	0
	Building	2
	Division	1
	Segment	1
	Work center	0
	Component	0
System function (material flow)	Manufacturing systems	0
	Logistics systems	0
	Building structure	4
	Process technology	0
	Building services	3
Energy form	Electricity	3
	Heat	0
	Excess heat	0
	Cold	0

4.8 Description of the Measure Implementation Support

The measure implementation support is required in order to provide relevant information on the assessment and realization of an EEM to the user. Defining structured categories enables a clear and understandable presentation of this information. It should be noted that approaches for structuring this general type of energy efficiency knowledge are hardly found in scientific literature. So far, only frameworks to structure application examples for EEMs are suggested (Roberts & Ball, 2014). Furthermore, action sheets are provided for the presentation of measures (for example in Seefeldt et al., 2006, pp. 234 ff.). However, these are developed for special applications, hence, present only partial information

on an EEM. Therefore, the qualitative, *i.e.*, textual, elements in practical guidebooks
are analyzed with regard to their information value and summarized in the following
categories (Bayerisches Landesamt für Umweltschutz, 2004; Bayerisches Landesamt
für Umwelt, 2009; Deutsche Bundesstiftung Umwelt, DBU, 2008; Deutsche Energie-
Agentur GmbH, dena, 2010a; Deutsche Energie-Agentur GmbH, dena, 2010b; Deutsche
Energie-Agentur GmbH, dena, 2010c; EnergieAgentur Nordrhein-Westfalen, 2010a;
EnergieAgentur Nordrhein-Westfalen, 2010b; EnergieAgentur Nordrhein-Westfalen,
2010c; Hessisches Ministerium für Wirtschaft, Verkehr und Landesentwicklung, 2009;
Lohre, Bernecker & Gotthardt, 2011):

- theoretical background,

- initial situation,

- targeted situation,

- benefit-effort ratio, and

- external information.

The *theoretical background* contains relevant terms and definitions and sets the basis
for understanding an EEM. The information in the category *initial situation* is used
to describe the situation before applying an EEM (*e.g.*, the causes for energy losses).
The *targeted situation* explains how an EEM works and on which principles it is based.
Furthermore, it specifies application areas and helps to put the focus on important
realization areas. Information on the *benefit-effort ratio* illustrates influences on expenses
and advantages when an EEM is applied. Finally, *external information* provides a link
to external information sources which might be helpful or required for the measure
implementation (*e.g.*, standards and laws).

Figure 28 shows the form to present the energy efficiency measure implementation
support (EEMIS). It should be noted that this framework can be applied in general for any
EEM, but not necessarily each EEM is assigned with information of each (sub-)category.
In the case of unused elements within the form, these may be removed.

The section *classification* contains selected classifying criteria of the EEM. The purpose
is to demonstrate the application area of the EEM. This helps users to ensure that
the measure is relevant for their planning case and may give ideas for extending its
application to other projects. The classification is based on objective, hierarchy, energy
form, product function, energy function, and other. The objective is usually to reduce
energy consumption, but as an extension of the method, it may also be used for measures
to reduce greenhouse gas emissions or energy costs. The criteria hierarchy, energy form,
product function, and energy function refer to the respective classifying criteria of the
EEM (see Section 4.7).

Energy efficiency measure		
Title of energy efficiency measure (EEM)		

Classification

Objective	Hierarchy	Energy form
Function (product)	Function (energy)	Other

Initial situation	**Targeted situation**
Description	Principle and variants
Description of initial situation (textual or graphical)	Explanation of how the EEM works (textual or graphical)
Cause	Application area
Explanation of the reasons for the deficits that the EEM addresses	Possible application areas for the EEM and assessment on focus areas (*i.e.*, where the EEM is especially recommended)
Relevance	Implementation
Information on the extent of the deficit in the initial situation	Information on measure implementation (*e.g.*, challenges, special requirements, extent of necessary changes)
Approaches to analyze initial situation	Employee involvement
Approaches to analyze the initial situation of the object system (*e.g.*, explanations on measurement methods)	Range and intensity of affected employees
Benchmark	Information need
Quantitative values on the energy consumption or losses in the initial situation (*e.g.*, practical examples, standard values)	Information that is helpful for assessing the applicability and impacts of the EEM

Benefit and effort

Energy savings	Investment	Pay-back time	Transfer time
Expected benefit	Explanations on expected benefits and effort (*e.g.*, influences on energy savings, theoretical calculation of benefit)		
Side effects	Effects on other objectives (*e.g.*, productivity) or other EEM		

Background	**External information sources**
Terms, definitions and theoretical explanations	Standardization
– Definitions of terms or units – Theoretical basics to understand the EEM	Relevant standards that foster or support the implementation of the EEM
Theoretical energy consumption	Legislation
Calculation of the energy consumption that is theoretically required to perform the task	Relevant laws and guidelines that foster or support the implementation of the EEM

Sources

[x] References to source documents	
Number	Date

Figure 28: Form Energy Efficiency Measure Implementation Support (EEMIS)

The category "Other" may be used for miscellaneous remarks, such as referring to the sector in the case of sector-specific measures.

Below the classification section is the main part of the sheet. It contrasts the initial situation, *i.e.*, before implementing a measure, with the targeted situation, *i.e.*, after implementation, in order to describe what needs to be done to realize an EEM. The section *initial situation* is explained using the following sub-categories: A general description explains the situation before a measure using text or diagrams. Furthermore, possible causes are explained (*e.g.*, physical explanation for energy losses, influencing factors that lead to a high energy consumption).

The relevance emphasizes the importance to take action since it points out the extent of the deficit (*e.g.*, amount of energy losses). Afterwards, approaches for analyzing the initial situation, *i.e.*, analysis and measurement methods and instruments, are described in order to create transparency on the specific situation of a user. Finally, the benchmark gives quantitative values on energy consumption or energy losses, with which the user may compare the specific system (*e.g.*, typical energy consumption of a similar system). The data source of the benchmark values may either be general rules (*e.g.*, engineering standards) or practical examples.

The section *targeted situation* gives information on the implementation of an EEM. At first, it demonstrates the principle, *i.e.*, how a measure works, and variants for implementing the measure using text or diagrams. Afterwards, the application area is described in more detail as compared to the classification section. This helps to put the focus on the measure realization (*e.g.*, necessary conditions, for which the EEM is especially promising). Furthermore, details on requirements and challenges for the implementation are mentioned (*e.g.*, required software or know-how). Due to the importance of considering a system as socio-technical, the employee involvement is described separately, *i.e.*, how employees are potentially affected by the EEM. This is helpful to prepare the communication concept before realizing a measure. Finally, the information need to assess the applicability or impact of the EEM is explained, which gives input to the definition of an implementation plan.

The section *benefit and effort* supports the technical and economic assessment of a measure. It refers to the classifying criteria energy savings, investment, pay-back time, and implementation time of an EEM (see Section 4.7). Furthermore, it contains a textual description of the expected benefit as well as on positive or negative side effects (*e.g.*, effect on productivity). Additionally, the side effects contain possible influences on other EEMs, which may either enhance or diminish each other. The textual description of the benefit may explain further parameters that influence the amount of the benefit in order to support the transfer of the general information to the specific use case.

Industrial application example				
Title of energy efficiency measure (EEM)				
Object area	Industrial sector		Location	
	Company name		Number of employees	
	Application area			
	Process description			
Implementation details	Initial situation			
	Measure description			
	Implementation year		Combination with other measures? ☐ No ☐ Yes:	
Benefit and effort	Assessment refers to	☐ Single measure	☐ Combination of measures	
	Energy savings	Energy cost savings	Investment	Pay-back time
	Explanations			
Sources	[x] Reference to source documents			
	Number	Date		

Figure 29: Form to present application example for energy efficiency measure

Further theoretical information is given in the section *background*, which contains terms, definitions, and theoretical calculations (*e.g.*, theoretically required energy to conduct a task). The reference to engineering standards and relevant laws and guidelines is presented in the section *external information sources*. Finally, the action sheet contains a reference to *sources*, which may be used for further reading.

Besides the EEMIS sheet, industrial application examples can be used to strengthen the practical background and to constitute a motivation for the realization of an EEM. Therefore, a second type of information sheet is conceived to present application examples of an EEM (Figure 29). The application examples are optional and linked to the EEM knowledge base. They may be based on literature or added after a successful realization.

The upper section *object area* specifies the background of the industrial example. It contains the definition of the company, sector, location, number of employees, application area, and process description. The main part of the sheet, the *implementation details*, is used to describe the initial situation and the realization of the EEM. The description may be in form of text and/or diagrams. It also specifies whether the industrial example contains a single measure or a combination of various measures. Below, the *benefit and effort* is quantified as far as possible. In some cases, the quantification of effects may refer to a combination of several measures, which is indicated on the sheet.

4.9 Matching Algorithm for Assigning Energy Efficiency Measures

The energy efficiency measure matching algorithm (EEMA) assigns the energy efficiency measures to the factory planning task based on the defined classifying criteria. In the following section, the notation of the input information is described, before the calculation steps are explained.

Overview and Input Information

In the previous sections, the domains of the method (see Section 4.2) are discussed regarding their structure and description. The goal of the method is to identify EEMs for a factory planning situation. This matching task is realized by an algorithm that takes into account the identified criteria to structure both the planning task and the EEMs.

The objective of the algorithm is to calculate a *suitability degree* for each EEM from the knowledge base that expresses whether the measure is appropriate for the defined planning task. The calculations are based on formal models of the planning situation and the EEMs, which are created by means of DMMs.

Each EEM is represented by a vector for each of the classifying criteria (see Tables 10 and 11 in Section 4.7). As such, the EEM i is described by the following vector with reference to the criterion ℓ with m_ℓ valid characteristics:

$$\vec{\alpha}_{\ell,i} = (\alpha_{\ell,i,1}, \ldots, \alpha_{\ell,i,m_\ell})^T, \text{with } \alpha_{\ell,i,j} \in \{0,1,2,3,4\}. \tag{4.14}$$

This means that the values $\alpha_{\ell,i,j}$ represent the relation between the EEM i and the j-th characteristic of criterion ℓ, whereof the number of characteristics is m_ℓ. The type of relations, *i.e.*, the meaning of the values between one and four, are explained in Section 4.7.

For example, the assessment is demonstrated for the measure "Place windows in order to maximize daylight" (number 125) in Table 12 in Section 4.7. The criterion hierarchical level ($\ell = 5$) is assessed positively for the characteristics building, division, and segment for this EEM. Hence, the corresponding vector is defined as

$$\vec{\alpha}_{5,125} = (\alpha_{5,125,1}, \ldots, \alpha_{5,125,6})^T = (0,2,1,1,0,0)^T. \tag{4.15}$$

Stacking these vectors for several EEMs as rows in a matrix results in the DMM between the EEMs and the criterion ℓ:

$$D_l = \begin{pmatrix} \alpha_{\ell,1,1} & \cdots & \alpha_{\ell,1,m_\ell} \\ \vdots & & \vdots \\ \alpha_{\ell,n,1} & \cdots & \alpha_{\ell,n,m_\ell} \end{pmatrix} \text{ with } \ell \in \{1, \ldots, L\}, \tag{4.16}$$

with n being the number of EEMs. Similarly, DMMs are created for all other classifying criteria of EEMs.

The planning situation is described as vector

$$\vec{\beta}_\ell = (\beta_{\ell,1}, \ldots, \beta_{\ell,m_\ell})^T, \text{with } \beta_{\ell,j} \in [0,1], \tag{4.17}$$

whereof the values $\beta_{\ell,j}$ represent the degree to which the j-th characteristic of the criterion ℓ applies to the planning situation. For the ease of modeling, it is possible to limit the range of eligible values to zero or one, *i.e.*, $\beta_{\ell,j} \in \{0,1\}$. It should be noted that the use of differing evaluations, *i.e.*, $\beta_{\ell,j} \in [0,1]$, does only affect the prioritization of EEMs but not their selection.

The matching algorithm between the planning task and the EEMs is conducted using the following four steps, whereof the last one is optional:

- pre-selection of required attributes,

- selecting suitable energy efficiency measures,

- prioritizing energy efficiency measures, and

- assigning user roles to energy efficiency measures.

These steps are explained in the following paragraphs.

Pre-Selection of Required Attributes

The first step is to eliminate EEMs that require mandatory attributes which are not fulfilled by the planning task. Thus, for each criterion ℓ, the rows of the corresponding DMM are examined for the value four. If this value applies, the vector of the planning task needs to be checked for this criterion. The EEM must not be selected if the criterion does not apply to the planning situation. Otherwise, even if it is fulfilled, it does not enter the calculation of the measure's suitability degree. In order to perform this step, the DMM D is changed into D^* containing the entries α^* for each criterion ℓ:

$$D_\ell^* = \begin{pmatrix} \alpha_{\ell,1,1}^* & \cdots & \alpha_{\ell,1,m_\ell}^* \\ \vdots & & \vdots \\ \alpha_{\ell,n,1}^* & \cdots & \alpha_{\ell,n,m_\ell}^* \end{pmatrix} \tag{4.18}$$

with

$$\alpha_{\ell,i,j}^* = \begin{cases} \alpha_{\ell,i,j} & \text{for } \alpha_{\ell,i,j} \neq 4, \\ 1 & \text{for } \alpha_{\ell,i,j} = 4 \text{ and } \beta_{\ell,j} > 0 \quad \text{with } i \in \{1, \ldots, n\} \text{ and } j \in \{1, \ldots, m_\ell\}, \\ 0 & \text{for } \alpha_{\ell,i,j} = 4 \text{ and } \beta_{\ell,j} = 0. \end{cases} \tag{4.19}$$

Selecting Suitable Energy Efficiency Measures

In the second step, each EEM is evaluated as being suitable or not based on the characteristics of the planning situation. Hence, this step includes a binary assessment. The result is represented by an assessment vector $\vec{\gamma}$. For each criterion ℓ, a sub-vector for the assessment is calculated by multiplying the DMM with the vector representing the planning situation:

$$\vec{\gamma}_\ell = D_\ell^* \cdot \vec{\beta}_\ell = (\gamma_{\ell,1}, \ldots, \gamma_{\ell,n})^T, \tag{4.20}$$

i.e., each entry represents the sub-assessment for an EEM. Afterwards, the binary assessment vector $\vec{\gamma}$ is calculated by:

$$\vec{\gamma} = (\gamma_1, \ldots, \gamma_n)^T \tag{4.21}$$

with

$$\gamma_i = \begin{cases} 0 & \text{for } \exists\, \ell \in \{1, \ldots, L\} : \gamma_{\ell,i} = 0, \\ 1 & \text{for } \gamma_{\ell,i} > 0\; \forall\, \ell \in \{1, \ldots, L\}. \end{cases} \tag{4.22}$$

This means that if any of the sub-assessment values is zero, the measure may not be applied. In the other case, *i.e.*, if all sub-assessment values are positive, the measure may be applied. A sub-assessment value equals zero if the characteristics of an EEM and the planning situation do not fit together. This is, for example, the case when an EEM may only be applied on the level of an entire factory but the planning task requests solutions on the level of a work center.

Prioritizing Energy Efficiency Measures

The final step is used to specify the measure evaluation, *i.e.*, the binary vector $\vec{\gamma}$ is changed into the vector $\vec{\gamma}^*$ to represent the suitability degree of each measure. This degree is only influenced by the relations "clearly defined" and "decisive" as defined in Section 4.7. The relation "not decisive" expresses the fact that an EEM may be suitable to all characteristics of a specific criterion, *i.e.*, the manifestation of this criterion does not matter to the EEM. In this case, the suitability degree is not influenced. The relation "required" has been accounted for in the first step of the matching algorithm.[13] The partial suitability degree, *i.e.*, for each criterion, is calculated by

$$\gamma_{\ell,i}^* = \begin{cases} \gamma_{\ell,i} & \text{for } \sum_{j=1}^{m_l} \alpha_{\ell,i,j} \neq m_\ell, \\ 0 & \text{for } \sum_{j=1}^{m_l} \alpha_{\ell,i,j} = m_\ell. \end{cases} \tag{4.23}$$

This means that the suitability degree is based on the sub-assessment values that evaluate the relation between an EEM and the planning task for each defined criterion. The condition excludes "not decisive" relations from the calculation. Aggregating these values into the suitability degree is normalized to

$$\gamma_i^* = \frac{\sum_{\ell=1}^{L} \gamma_{\ell,i}^*}{3 \cdot \sum_{\ell=1}^{L} \sum_{j=1}^{m_\ell} \beta_{\ell,j}}, \tag{4.24}$$

---

[13] In the first step of the algorithm, the "required" relations are transformed to "not decisive" relations in order to simplify the further calculations.

whereof the denominator represents the optimal suitability. This is the case when each selected criterion is fulfilled with a greatly important corresponding characteristic (which is quantified by the value 3).

Assigning User Roles to Energy Efficiency Measures

Having identified suitable EEMs, these may be assigned to the roles of the planning participants. The difficulty of this assignment is due to the absence of direct linking between the EEMs and the user roles in the knowledge base (see Section 4.7). Hence, the assignment between EEMs and actors is conducted indirectly through the energy efficiency influential parameters. Therefore, the influence of an actor on the energy efficiency influential parameters needs to be specified as part of the actor description (see Section 4.6).

Among all criteria of an EEM, $\ell \in \{1, \ldots, L\}$, the energy efficiency influential parameters represent the criterion ℓ' of an EEM. A special feature of this criterion is that the relations between it and the EEM are either clearly defined or decisive. This means than an EEM may affect one or more energy efficiency influential parameters but not all of them. The relevant DMM is represented by

$$
D_{\ell'} = \begin{pmatrix} \alpha_{\ell',1,1} & \cdots & \alpha_{\ell',1,m_{\ell'}} \\ \vdots & & \vdots \\ \alpha_{\ell',n,1} & \cdots & \alpha_{\ell',n,m_{\ell'}} \end{pmatrix}.
\tag{4.25}
$$

Furthermore, the influences of actors on the parameters are modeled by the influence matrix I. This matrix shows the effect of a role on parameters with reference to the planning phases. The influence matrix may formally be represented as

$$
I_r = \begin{pmatrix} \omega_{r,1,1} & \cdots & \omega_{r,1,m_{\widehat{\ell}}} \\ \vdots & & \vdots \\ \omega_{r,m_{\ell'},1} & \cdots & \omega_{r,m_{\ell'},m_{\widehat{\ell}}} \end{pmatrix}
\tag{4.26}
$$

for a specific role $r \in \{1, \ldots, R\}$ with the element $\omega_{r,k,j}$ representing whether actor r has an influence on the parameter k in the planning phase j. Since the influence matrix is affected by the planning phases, they need to be considered for the role assignment. The planning phase is understood as the criterion $\widehat{\ell}$ of an EEM. These relations are represented by the DMM $D_{\widehat{\ell}}$.

The assignment between EEMs and actors is performed by generating a binary measure-actor-matrix

$$M = \begin{pmatrix} \mu_{1,1} & \cdots & \mu_{1,R} \\ \vdots & & \vdots \\ \mu_{n,1} & \cdots & \mu_{n,R} \end{pmatrix}, \tag{4.27}$$

which values indicate whether an actor $r \in \{1, \ldots, R\}$ is assigned to a measure $i \in \{1, \ldots, n\}$. Within this matrix, the rows represent the EEMs and the columns represent the actors. The matrix entries are calculated by

$$\mu_{i,r} = \begin{cases} 1 & \text{for } \exists\, j \in \{1, \ldots, m_{\hat{\ell}}\} \wedge k \in \{1, \ldots, m_{\ell'}\} : \alpha_{\hat{\ell},i,j} > 0, \alpha_{\ell',i,k} > 0, \omega_{r,k,j} > 0, \\ 0 & \text{else.} \end{cases}$$

$$\tag{4.28}$$

This means that the assignment value $\mu_{i,r}$ between an EEM i and a role r is true if the role r influences the parameter k in the planning phase j, the EEM i influences the parameter k, and may be applied in planning phase j.

Algorithm Results

The calculation results of EEMA include the suitability degree $\vec{\gamma}^*$ and the measure-actor-matrix M. The suitability degree is represented as a vector with the dimension n, i.e., the number of EEM in the knowledge base. The values of this vector represent the degree, to which the respective measure fits the planning task. The measure-actor-matrix contains the assignment between EEMs and actors. This matrix contains values of zero or one, whereof the value one means that the respective actor should be considered for the respective measure. Therefore, the results can be summarized as a list of measures that are sorted depending on their applicability for the planning task and assigned to relevant actors (see Table 13).

Table 13: Exemplary list of energy efficiency measures and assignment to actors as algorithm results

No.	Measure	Suitability degree	Actor assignment			
			Actor 1	Actor 2	Actor 3	...
1	...	69 %	X	X	-	...
2	...	23 %	-	X	X	...
3	...	10 %	X	-	-	...

4.10 Procedure Model for Method Application

The procedure model demonstrates how to apply the method to identify EEMs for a specific factory planning task. Its approach is based on the general systems engineering problem-solving paradigm (Section 3.2.2). Figure 30 shows the steps of the procedure in SADT notation (see Section 3.3.2). The steps are explained in the following paragraphs.

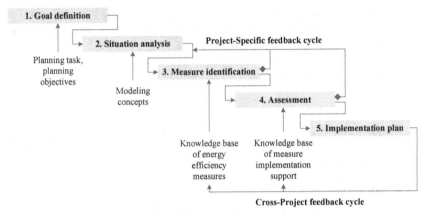

Figure 30: Procedure model for the methodical approach

Goal Definition

The first step is to define the goal for applying the method. The mainly supported objective is to reduce energy consumption. This may be achieved by reducing the power need or by reducing the period of time, during which a power demand occurs. As an extension of the suggested methodical approach, other objectives, such as the reduction of greenhouse gases, are possible. Furthermore, the application area is selected, *i.e.*, the boundaries of the system are defined. The application area may be the entire factory system or any subsystem of the factory. In the case of a brownfield project, the focus of the method's application may be based on areas that cause a high energy consumption. If quantitative values are available on the factory's energy consumption, several existing approaches may be considered to select an object area to focus on. For example, a classification of energy consumers into a portfolio according to their nominal value and operating time is a suitable approach (Thiede, Bogdanski & Herrmann, 2012, p. 31).

The level of detail for the method's application is specified according to the project purpose. This is required as a basis to determine the subsequent steps. On the one hand, the method might be used to conceive general principles to reduce energy consumption in a factory of a specific industrial sector (rough level of detail). On the other hand, a high level of detail may be applied in the cases where detailed insights on the technical parameters of a system are analyzed. Moreover, it needs to be decided whether an assignment between EEMs and actors is desired.

Situation Analysis

The second step is the qualitative analysis of the situation, *i.e.*, the planning task. The analysis contains the description of the modeling domains:

– object system,

– energy efficiency influential parameters,

– project characteristics, and

– actors.

The first substep is to describe the technical object system, which contains general characteristics of the factory as well as the hierarchical and functional structure of the considered system (Section 4.3). The purpose is to define subsystems, elements, and processes in the selected factory system in order to generate an understanding of the object area. For each relevant subsystem, the input and output energy flows are defined. Additionally, the operational states are defined on system and subsystem levels. Afterwards, the energy efficiency influential parameters are identified for each subsystem. This means, variables are determined that influence the intensity of an energy flow (Section 4.4). The effect of any parameter may be limited to a specific operational state or may refer to the total system energy consumption. With specifying the energy efficiency influential parameters, the functional model of the object system is extended.

The next substep is to define the project characteristics (Section 4.5). This contains both parameters of the factory planning project (life cycle phase of area, building, equipment, and product, planning phase) and of the energy efficiency project (project size). Finally, the relevant actors are described. For the selection of actors, both planning participants and persons affected from the planned result may be considered. This substep includes the description of actors, their organizational units, and tasks. Furthermore, a matrix is used to represent the influence of the actors on relevant energy efficiency influential parameters, whereof this influence may vary between life cycle phases and planning phases (see Section 4.6).

Identification of Energy Efficiency Measures

Based on the situation analysis, suitable EEMs are selected from the knowledge base. The assignment is based on the classifying criteria as defined in Section 4.7. The developed criteria contain both identification criteria and assessment criteria. The identification criteria describe the applicability of an EEM, *i.e.*, whether a measure may be applied in the project. The assessment criteria represent the effort and benefit in terms of energy savings, investment, pay-back time, and implementation time. It is not necessary to determine each criterion to perform the assignment task. However, with a limited specification of parameters, the results may yield a high number of EEM on a general level. The assignment is realized by the EEMA algorithm (Section 4.9).

Assessment and Selection of Energy Efficiency Measures

In the fourth step, the EEMs are reviewed by the planning participants. The purpose is to qualitatively assess the applicability and usefulness of the measures for the project. This requires defining assessment criteria and evaluating the alternatives towards these criteria. Methods to support this assessment are the value benefit analysis or the point rating method (see Section 3.2.3).

The information on the EEM is provided by the measure implementation support as described in Section 4.8. It contains both a general action sheet of an EEM and industrial examples that demonstrate the implementation of a measure. Both information sheets are connected to the knowledge base of EEMs. It should be noted that the information provided by the measure implementation support does not quantitatively refer to the project, *i.e.*, the benefit of an EEM in terms of energy savings as described on the EEMIS sheet is estimated based on other case studies and may differ from the specific case. When selecting EEMs for implementation, interdependencies between several measures need to be considered. Therefore, the EEMIS sheet contains a section on side effects and effects on other EEMs. Furthermore, secondary effects of measures that arise from the interaction between object systems should be analyzed.

Development of an Implementation Plan

The last step of the procedure is to develop a plan for the implementation of the selected EEMs. Since the information provided by the methodical approach is in a qualitative manner, this may include more detailed analyses in order to quantitatively assess the benefit and effort of a measure (see Section 4.1). The implementation plan should contain responsibilities, deadlines for the realization, and measurable goals (VDI 4070, Part 1, p. 13). Furthermore, methods to examine a measure's effect should be defined (*e.g.*, energy measurement after a reasonable period of time). Developing an implementation plan for concrete measures may be guided by researching funding opportunities (*e.g.*, from federal ministries). The status of realization should be documented in the implementation plan in order to support a continuous review and improvement.

Project-Specific Feedback Cycle

The project-specific feedback cycle may be entered at each step of the procedure after the identification of EEMs. It allows going back to the definition of the system boundaries and specification of subsystems in case the resulting EEMs do not fulfill the implicit requirements. This may be the case if the specification of the situation is too general, which leads to more generic measures and may reduce the transferability to a specific project. On the contrary, it is possible that the number of identified measures is too small due to restrictions in the definition of the project. For example, planning participants may attempt to identify measures that are applicable with a low budget in a short period of time, even though there are hardly possibilities to do so in the selected object area. In this case, restrictions need to be relaxed in order to generate an appropriate number of resulting measures.

Cross-Project Feedback Cycle

The cross-project feedback cycle is implemented in order to increase the knowledge base of the method. The feedback should be performed both after defining the implementation plan and again after a reasonable period of time. The prerequisite for applying the method is a collection of EEMs and respective implementation support. Enterprises may gain experience on EEMs and may want to add them to the knowledge base in a long-term application. This may include, among others, challenges with the implementation, newly identified application areas or measure variants. Furthermore, the lessons learned from implementing a measure may be added as industrial examples. While the implementation support can be added by changing an existing or adding a separate document, an additional EEM needs to be classified according to the criteria as described in Section 4.7.

4.11 Results of the Method Development

The identification of energy efficiency measures for factory systems is a complicated task due to the complexity of factory systems and the variety of principally possible energy efficiency measures. A systematic procedure for identifying energy efficiency measures needs to consider the entirety of objects in a factory system while posing little requirements on data availability. This is especially due to the purpose of applying the methodical approach during early factory planning phases, in which quantitative data might not be available.

The developed method aims at assigning energy efficiency measures to a factory planning task based on a qualitative analysis. Furthermore, the energy efficiency measures are complemented by information on the measures' implementation. The major components of the method are description concepts for both the factory planning task and the energy efficiency knowledge, an assignment algorithm and a corresponding procedure model. The socio-technical representation of the factory planning task is realized by describing the domains object system, energy efficiency influential parameters, project characteristics, and actors.

The object system represents the technical resources in a factory and describes them in terms of their hierarchy and function. When specifying the function of a factory subsystem, both the material flow (*e.g.*, manufacturing, logistics) and the energy flow (*e.g.*, distribution, usage) are considered. The input and output of energy flows into and out of a system need to be determined. Finally, different operational states of a system are distinguished (*e.g.*, operating, standby). The energy efficiency influential parameters describe variables that affect these flows (*e.g.*, the mass of a conveyor influences its energy consumption during operation).

The social aspects of the factory system are part of the modeling domains project characteristics and actor. The characteristics of the factory planning project and the energy efficiency project may limit the applicability of energy efficiency measures. Hence, the project is assigned to the factory life cycle and to the factory planning phase. Furthermore, the scope of the energy efficiency project is represented in terms of costs and time. The actors are described by their tasks towards the object system and their effects on the energy efficiency influential parameters.

The energy efficiency measures are structured according to criteria that are relevant for identification and assessment. These criteria are the basis to conduct the assignment between a planning task and the corresponding energy efficiency measures. The knowledge base on energy efficiency measures contains the measures themselves and the assessed relation between the measure and each of the classifying criteria (*e.g.*, evaluated applicability of an energy efficiency measure during a specific factory planning phase). The measure implementation support contains information that is relevant for the measure's realization. It is provided in form of the standardized EEMIS sheet that describes the initial and targeted situation as well as benefit and effort of a measure. Furthermore, industrial examples are presented. Planning participants are assisted in evaluating and selecting appropriate measures by means of the measure implementation support.

The matching algorithm EEMA performs the assignment of energy efficiency measures to the factory planning task. For this purpose, the measures are formally modeled with domain mapping matrices that represent the previously defined classifying criteria. The analysis of the planning task leads to a formal model, which is mapped to the EEM in the knowledge base. The result of the algorithm is a list of appropriate measures including their assignment to the respective actors. The list is prioritized with the measures' suitability degree, *i.e.*, a percentage value that describes the fit between a measure and the defined factory planning task.

The procedure model describes the application of the method for a specific factory planning task. The procedure comprises the steps goal definition, situation analysis, identification of energy efficiency measures, assessment and selection, and development of an implementation plan. This approach systematically guides factory planning participants through the identification of energy efficiency measures. An iterative feedback allows participants to revise the identified measures. Furthermore, gained knowledge during factory planning projects may be incorporated in the knowledge base since the approach allows to extend both the energy efficiency measures and the measure implementation support.

In summary, the developed method provides an approach to identify energy efficiency measures for factory systems. It helps to create transparency on the possibilities to influence the energy consumption of a factory and the roles of various actors to perform this task. Furthermore, information on measures is provided that supports factory planning participants in generating energy-efficient planning solutions.

5 Validation

In chapter five, the developed method is validated in order to prove relevance and usefulness with regard to the applied research process of the thesis. At first, suitable validation objectives and methods are identified as part of the validation concept. Based on that, the relevance of the method is validated by assessing the determined requirements. Afterwards, a prototype is generated that implements the energy efficiency knowledge base and the matching algorithm. The focus of the chapter are two comprehensive case studies with different application areas. Finally, a review of the results summarizes the validation findings.

5.1 Validation Concept

The purpose of the validation is to review the developed method as part of the application-oriented research process. Validation means to determine whether a system satisfies specified requirements (IEEE Dictionary). These requirements are deduced from an intended use and application (DIN EN ISO 9000, p. 50). As a subset of validation, testing may be performed in order to determine whether a system functions properly (Engel, 2010, p. 66). Thus, the validation within this thesis follows the purpose to assess whether the intended goals are achieved.

It should be noted that the validation is an ongoing process during system development, which means a continuous reflection on the fulfillment of defined requirements (Chan, 2014, p. 10). Hence, the research process to develop the methodical approach has been carried out while continuously checking the fulfillment of requirements.[14]

Validation strategies are distinguished according to whether they validate against the research gap or the "real world" (Riege, Saat & Bucher, 2009, p. 75). The latter strategy focuses on the practical usefulness of a method. A validation activity comprises an objective, a description (*i.e.*, purpose and implementation) and the applied validation method (Engel, 2010, p. 65). Following the goal of the thesis and the application-oriented research process, the validation objectives are relevance and usefulness (Ulrich, 1984, p. 175). The relevance describes the fulfillment of methodical requirements. The usefulness focuses on the practical implementation of the method.

Potential validation methods include prototypes, characteristic-based comparisons, surveys, simulations, and case studies (Bortz & Döring, 2006; Karlsson, 2009; Riege, Saat & Bucher, 2009). The choice of suitable validation methods depends on the validation

---

[14]This includes the presentation of the method or parts of it at both national and international conferences, such as the Magdeburg Logistics Days, the International Conference on Flexible Automation and Intelligent Manufacturing (FAIM), the International Conference on Production Research (ICPR), the CIRP Conference on Manufacturing Systems (CMS), and the International Conference on Industrial Engineering and Operations Management (IEOM).

strategy, objectives, and the conditions of the research process, such as availability of data (Riege, Saat & Bucher, 2009, p. 82).

The relevance of the developed method is validated against the research gap. Therefore, a characteristic-based comparison is used to check the fulfillment of methodical requirements. These requirements are identical to the criteria to assess the state of the art (Section 3.5.5).

The validation of the usefulness is supposed to consider the requirements of an industrial application. Case studies enable this goal, since they are useful to capture the complexity of a case in detail (Bortz & Döring, 2006, p. 110). Hence, the usage of case studies represents an opportunity to validate the practicability and functionality of the developed method. Whether a case study is representative, depends on the type of the case and the homogeneity of the basic population (Bortz & Döring, 2006, p. 323). Factory planning projects are characterized by their complexity with regard to the factory system and the interdisciplinarity of the project team. Due to this complexity, two comprehensive case studies are carried out in order to validate the usefulness of the method. Since conducting use cases requires energy efficiency knowledge, a prototype is realized including an initial knowledge base and an implementation of the EEMA algorithm. Table 14 summarizes the validation concept.

Table 14: Validation concept

Objective	Validation method	Description
Relevance	Characteristic-based comparison	Assessment of the evaluation criteria for the state of the art (Section 5.2)
Usefulness	Prototype	Prototypical realization of the knowledge base and the matching algorithm (Section 5.3)
Usefulness	Case study	Planning of welding processes (Section 5.4)
Usefulness	Case study	Planning of logistics systems (Section 5.5)

5.2 Review of Methodical Requirements

The purpose of this section is to review the fulfillment of the requirements for the developed method. Hence, the method is validated using the same assessment criteria as for the state of the art on approaches to increase energy efficiency in factories (see Section 3.5.5). These criteria are mainly identified based on current barriers for the implementation of EEMs in industrial enterprises (see Section 2.3) and against the background of the application in factory planning (see Section 3.4).

Practicability of Data Acquisition

The necessary energy data for applying a method poses a major barrier for industrial application. The analysis of the state of the art shows that existing methods focus on the quantitative analysis of a system. Hence, the effort to gather the required data tends to be high. The requirement of quantitative data may include either load profiles that are acquired through energy measurements or a quantitative energy demand (*e.g.*, provided through data sheets from the equipment manufacturer).

A low energy data requirement, however, is set by methods that only need qualitative information. The data and information requirements of the developed methodical approach comprise the technical systems including their energy flows and influential parameters, project requirements, and actors. This information is purely qualitative, which means that the practicability of data acquisition is high.

Systematic Optimization Procedure

Since existing methods to increase energy efficiency focus on analysis rather than on the optimization of a system, the identification of improvement measures is less supported. While some methods create transparency on the energy consumption of a system, they do not provide techniques for deducing corresponding measures. Other instruments provide guidelines on energy efficiency principles. Yet, the transfer to the individual situation needs to be conducted by the planning participants.

On the contrary, the developed methodical approach describes a procedure to guide the identification of EEMs that are tailored to the individual planning task. A major advantage of the method lies in the quick identification of solution approaches by the developed algorithm.

Completeness of Objects

EEMs in factories should address all types of object systems in order to achieve a significant effect. However, several existing methods focus on partial systems, such as manufacturing processes. The developed methodical approach considers the variety of factory systems. In the first step of the procedure, the system is defined, which may include an entire factory or selected partial systems with regard to the preference of the user. The methodical procedure does not differ depending on the selected system. Hence, all objects within a factory may be analyzed by using this method.

Completeness of Energy Flows

When addressing energy efficiency, the majority of approaches focuses on reducing electricity consumption since it is a widely applied energy source in industry. However, the main energy form may be different depending on the industrial sector (*e.g.*, process heat). Hence, it is important to allow the consideration of various energy flows. As part of the methodical procedure, the relevant object systems are described including their energy flows. This means that the method does not include a restriction towards the energy flows that may be addressed.

Range of Measures

The usefulness of the resulting EEMs depends on whether they are limited to a specific range. For example, in simulation studies, the resulting measures are usually restricted to manipulating the parameters that enter a simulation model (*e.g.*, reduce cycle times). The developed method assigns EEMs to the individual planning task, whereof the measures are part of a comprehensive knowledge base. Since this knowledge base may be continuously supplemented, the resulting measures are not restricted to a specific type.

Specificity of Measures

Another important aspect for supporting factory planning participants is the need to transfer the results of a method to their individual planning situation. The effort for this transfer task highly depends on the specificity of the resulting measures. As such, a general principle (*e.g.*, reduce process losses) requires a high transfer effort. A higher specification is achieved when the resulting measures are described with any reference to the relevant factory system. The developed method fulfills this criterion since the deduction of EEMs is tailored to the specific planning task, *i.e.*, through matching the classifying criteria of the planning task with the criteria of EEMs (see Section 4.7 for the criteria of EEMs and Section 4.9 for the matching algorithm).

Applicability in Factory Planning

The majority of existing approaches focus on an application during factory management, *e.g.*, due to the required quantitative data. This data might not be available during early planning phases. However, the influence on the energy consumption of a factory is especially important in these early phases. The developed method is tailored to the requirements of factory planning, especially through considering the different life cycle stages of a factory and the interdisciplinary actors in factory planning tasks. Hence, the method is applicable in factory planning.

Transferability

The transferability of a method into industrial application depends on its transparency and reproducibility of results. If a method depends on expert knowledge, it tends to be less conducted by industrial practitioners. Hence, a clear description of the method and documentation of each step is important to reduce implementation barriers. The developed procedure model serves this task. The identification of EEMs is performed by clearly defined criteria and realized through an algorithm. The prototype demonstrates an initial implementation of the knowledge base of EEMs and the EEMA algorithm. By this, the transfer of the method into industrial application is additionally supported.

The review of requirements shows that the criteria are entirely fulfilled by the developed method. Hence, the method development has achieved the goal of a systematic procedure that suggests solution approaches for energy efficiency optimization during factory planning.

5.3 Prototype

A prototype for the methodical approach is created based on Excel.[15] It contains an initial knowledge base of both EEMs and EEMIS as well as the matching algorithm (see Section 4.9). The knowledge base consists of 200 measures that result from both literature research and experience of the author through conducted energy efficiency projects.[16] It is structured as a table with a row for each EEM. The columns contain the classifying criteria as defined in Section 4.7. The entries represent the values of the DMMs between the EEM and the criteria (see Section 4.9). Figure 31 shows an extract of the knowledge base.

No.	Measure	Hierarchical level					
		Component	Work Center	Segment	Division	Building	Factory
54	Regular maintenance of drives	3	0	0	0	0	0
55	Parameterization of frequency converters	3	0	0	0	0	0
56	Use pole switchable motors	3	0	0	0	0	0
57	Optimize energy demand during movement breaks	2	2	0	0	0	0
58	Use efficient servocontrollers	3	0	0	0	0	0
59	Reduce cable length to motors	3	0	0	0	0	0
60	Recover braking energy of drives	3	0	0	0	0	0
61	Use direct drives	3	0	0	0	0	0
62	Use gear drives with high efficiency	3	0	0	0	0	0
63	Use electric motors with softstarter functionality	3	0	0	0	0	0
64	Reduce illumination through removing lamps	0	3	0	0	0	0

Figure 31: Exemplary extract of the prototypical knowledge base

---

[15] The software version is Microsoft Excel 2013 with Microsoft Visual Basic for Applications 7.1.

[16] This includes working in the cluster of excellence "Energy-efficient product and process innovations in production engineering" and several industrial projects in the context of energy-efficient factories in automotive industry. During the thesis work, the author has successfully completed the certificate program European Energy Manager (EUREM). Furthermore, the author has offered workshops on energy-efficient factories for industrial participants from automotive industry.

The method's application is guided by a graphical user interface and presents the results in form of a table. The inquiry forms support the step situation analysis by providing the relevant criteria to describe the planning task (Figures 32 and 33). The user's task is to select the criteria to be considered and their applying characteristics.

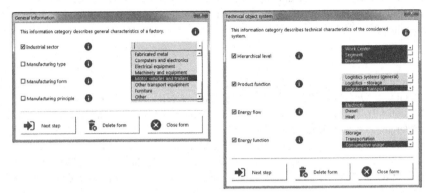

Figure 32: Prototypical user interface for general, hierarchical, and functional description of the factory system[17]

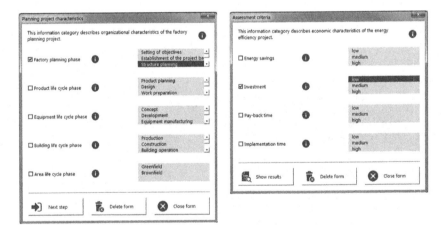

Figure 33: Prototypical user interface for the description of project characteristics and assessment criteria

At first, general information on the factory system is specified (see Section 4.3.1). This includes the industrial sector, manufacturing type, manufacturing form, and manufacturing principle (see Figure 32, left). The next step is to provide the hierarchical and functional

---

[17] Icons made by Freepik, Dave Gandy, and Eleonor Wang from www.flaticon.com

description of the object system according to Sections 4.3.2 and 4.3.3 (see Figure 32, right). Afterwards, the project characteristics in terms of planning phases and life cycle phases (Section 4.5) are determined (see Figure 33, left). Finally, the assessment criteria of EEMs, *i.e.*, the scope of the energy efficiency project are determined (Sections 4.5 and 4.7). This means to provide the expectations on energy savings, investment, pay-back time, and implementation time (see Figure 33, right).

Having specified the input for these substeps of the situation analysis, the input is automatically converted to the corresponding description vector of the planning task (see Section 4.9). This information is enclosed in a separate worksheet of the document, hence, the user may verify the entries or easily change them. The result of the EEMA algorithm is a prioritized list of applicable EEMs showing their suitability degree (see Figure 34). This refers to the step "Identification of energy efficiency measures" of the procedure model, yet only considering the first three steps of the matching algorithm (see Section 4.9).

No.	Measure	Industrial sector	Manufacturing type	Manufacturing form	Manufacturing principle	Hierarchical level	Product function	Energy form	Energy function	Planning phase	Product life cycle	Equipment life cycle	Building life cycle	Area life cycle	Energy savings	Investment	Pay-back time	Implementation time	Suitability degree
2	Reduce masses to be transported	0	0	0	0	3	3	3	3	5	0	8	0	0	3	0	0	3	57%
6	Increase ratio between load capacity and total mass	0	0	0	0	3	3	3	3	6	0	8	0	0	3	0	0	0	54%
25	Reduce areas for transport and providing material	0	0	0	0	3	2	3	3	3	0	8	0	0	3	3	0	0	52%
9	Increase utilization rate of transport containers	0	0	0	0	4	3	3	3	3	0	8	0	0	3	0	0	0	50%
19	Reduce mass of containers and packaging material	0	0	0	0	3	3	3	3	4	0	8	0	0	0	0	0	0	44%
8	Use vehicle fleet management	0	0	0	0	3	3	3	3	0	8	0	0	0	0	0	0	0	43%

Figure 34: Exemplary result of the measure identification in the prototype

The forth optional step contains the assignment between EEMs and relevant actors. This requires to specify the influence matrix, *i.e.*, the effect of a role on the energy efficiency influential parameters, which may be conducted by the user in a separate worksheet. Therefore, the user may enter the number of relevant roles, which is the basis to generate empty matrices considering the number of considered planning phases.[18] The worksheet to enter the influence matrix is depicted in Figure 35. The task of the user is to first enter the energy efficiency influential parameters as identified in the step "Situation analysis" of the procedure model. Afterwards, the influence is marked for each role, parameter, and phase by using the values 1 or 0. If the influence matrix is specified, the resulting list of EEMs is extended. This means that for each measure the relevant actors are marked in the measure-actor-matrix (see Figure 36).

---

[18] The influence of roles may vary depending on life cycle or planning phases (see Section 4.6).

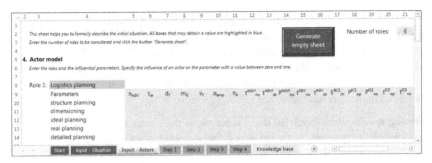

Figure 35: Worksheet to describe the actors' influences

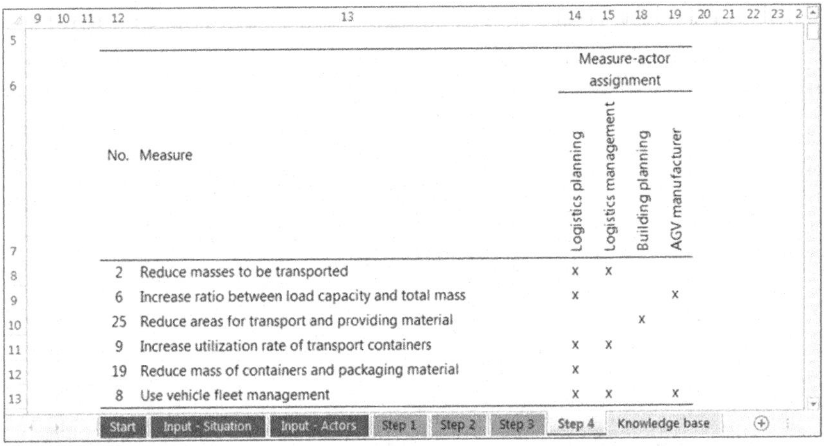

Figure 36: Exemplary result of the measure-actor matrix in the prototype

The step "Assessment and selection of energy efficiency measures" of the procedure model is supported by the EEMIS sheets and industrial examples (see Section 4.8). The prototype contains links to these documents in form of separate PDF files.

As a summary, the prototype supports the identification of EEMs based on the specification of a factory planning task. Therefore, the steps of the procedure model are implemented. The situation analysis is supported by graphical user interfaces that allow to specify the domains object system, project characteristics, and actor. The implemented algorithm identifies suitable EEMs from the knowledge base, which initially contains 200 measures. The result is depicted in form of a list of the identified EEMs including their suitability degree that represents the fit to the planning task. Optionally, the influence of actors may be specified. In this case, the list of results contains the assignment of relevant actors. The prototype serves as a basis to conduct the following case studies.

5.4 Case Study 1: Planning of Welding Processes

The first case study addresses the identification of energy saving potentials when planning welding processes.[19] Following the procedure model in Section 4.10, the steps goal definition, situation analysis, and identification of energy efficiency measures are performed.

5.4.1 Goal Definition

The task of this case study is to plan new welding equipment for an existing job shop including the preparation of ramp-up and operation. The case study serves as an example for identifying general measures on a low level of detail in an early planning stage. Therefore, the step to assess and select measures for implementation is not conducted. Prior to the method's application, basics on welding processes are explained to support the understanding for the case study.

5.4.2 Basics on Welding Processes

In general, a joining process is defined as bringing together two or more parts that have a geometrically defined form (DIN 8580, p. 5). Joining processes may use several mechanisms to create the joint: A form closure is based on the geometry of the parts, whereas a force closure creates the joint by a force. A material closure joins objects through a material connection (*e.g.*, welding).

Welding is the process to fuse together two materials either by heating, with or without the application of pressure, or by the application of pressure alone, and with or without the use of filler materials (AWS A3.0-94, p. 38). Two general technologies are represented by solid state welding and fusion welding. In solid state welding, the process temperature stays below the melting temperature of the welded material. Joining is realized by a plastic deformation through applying pressure. Solid state welding usually takes place without filler material (Fahrenwaldt, Schuler & Twrdek, 2014, p. 3). Fusion welding processes are realized by melting the joining surface and re-solidifying the material afterwards.

Moreover, welding processes may be characterized depending on the active energy carrier, *i.e.*, the source that provides energy to create the joint. This includes solid body, liquid, gas, electrical discharge, radiation, movement of a mass, and electric current (DIN EN 14610, p. 8).

---

[19]The case study is based on a project that was conducted as part of a research stay at the Laser-assisted Multi-scale Manufacturing Laboratory at the University of Wisconsin-Madison in 2014. This project dealt with comparing the energy consumption between gas-metal arc welding and friction stir welding (Shrivastava, Krones & Pfefferkorn, 2015).

An important group of fusion welding methods is arc welding, in which the melting heat is provided by electrical energy. The process includes an electrode, made out of rod or wire, and an electrical arc between the electrode and the workpiece (Hutchins, Robinson & Dornfeld, 2013, p. 538). Processes within this group are explained in the following (DIN EN 14610, pp. 59 ff.): *Manual metal-arc welding* (MMAW) is operated with a melting rod electrode that is manually fed. The electrode is covered, which protects the weld pool. *Submerged arc welding* (SMAW) is operated with a wire electrode and the arc is enveloped by slag which fuses from granular flux. The flux is deposited loosely in the weld area.

Gas-shielded welding is a group of processes, which use gas from an external source to shield the weld pool. The processes in this subgroup are distinguished according to the type of gas (active or inert) and the type of electrode (consumable or non-consumable). Table 15 summarizes the processes in gas-shielded metal-arc welding. Metal inert gas welding (MIG) and metal active gas welding (MAG) may be summarized to gas-metal arc welding (GMAW). Similarly, gas-tungsten arc welding (GTAW) comprises both tungsten inert gas welding (TIG) and tungsten active gas welding (TAG). (DIN EN 14610, pp. 59 ff.)

Table 15: Overview on gas-shielded metal-arc welding processes

| | | Type of electrode | |
		Consumable	Non-consumable
Type of gas	Inert	Metal inert gas welding (MIG)	Tungsten inert gas welding (TIG)
	Active	Metal active gas welding (MAG)	Tungsten active gas welding (TAG)

Plasma arc welding is similar to tungsten inert gas welding, whereof the welding arc is constricted by a plasma that is forced through a copper nozzle. Plasma is the fourth physical state (besides solid, liquid, and gaseous) and may be described as a thermally highly heated and electrically conductive gas. Plasma arc welding is characterized by higher temperatures, a smaller opening angle and a higher power density. *Laser welding* is a fusion welding process using the energy of a laser, *i.e.*, a coherent beam of light, to melt the material. Laser welding may be combined with arc welding processes – *laser-hybrid arc welding* (LHAW) – whereof an example is laser gas tungsten arc welding (Laser-GTA). (DIN EN 14610, pp. 59 ff.)

As an example for solid state welding, friction stir welding (FSW) is explained in the following (Shrivastava, Krones & Pfefferkorn, 2015, p. 159): The workpieces are plastically deformed and mechanically intermixed at elevated temperatures. The FSW tool is rotating and plunged into the workpiece. Once the pin of the tool is completely inserted into the workpiece, the tool is traversed along the welding path and finally retracted at the end of the weld. The weld energy is generated due to friction between the tool and the workpiece and by the plastic deformation of the material, *i.e.*, the stirring. Since the welding process takes place below the solidus temperature, this process is suitable for joining dissimilar materials, *e.g.*, aluminum and magnesium.

The fundamental functionality of welding processes and further terms are explained for the example of welding a butt joint with GMAW, *i.e.*, between two workpieces that are aligned in the same plane (Figure 37). Since the process applies additional material through the consumable electrode, the parts are prepared with a groove before welding. Typical groove geometries contain V- or U-shape and may be applied on one side (single-groove weld) or both sides of the surface to be joined (double-groove weld).

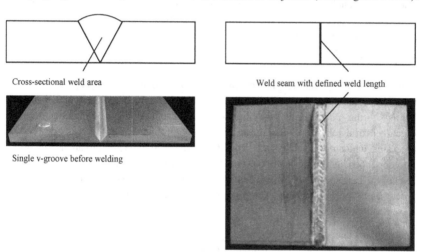

Cross-sectional weld area

Single v-groove before welding

Weld seam with defined weld length

Figure 37: Exemplary parts geometry for creating a butt joint with gas-metal arc welding

During the GMAW process, the consumable electrode is automatically fed through the welding nozzle with a defined velocity (wire-feed speed), which is determined depending on the material and the weld thickness. According to the feed rate and the wire diameter, there is a resulting amount of additional material which is deposited into the weld zone (deposition rate). The wire-feed speed is directly related to the welding current, *i.e.*, the electrical amperage of the welding equipment. Moving the nozzle along the weld seam is performed in a defined velocity, the welding speed or arc travel speed.

The generated heat locally melts the base material and the electrode (weld pool). Afterwards, the material solidifies as weld metal. Besides the weld metal area, a portion of the base metal changes its mechanical properties due to the welding heat (heat-affected zone).

The quality of a weld may be assessed by both non-destructive and destructive testing methods, including the criteria outer and inner weld defects, imperfections in welding seam geometry, tensile strength, bending strength, and hardness (DIN EN ISO 5817, pp. 9 ff.; DIN EN ISO 15614, p. 13). A quality indicator is the *joint efficiency*, *i.e.*, the ratio between the tensile strength of the welded structure and the strength of the base material (Mishra, De & Kumar, 2014, p. 177). For the example of AlMg1SiCu alloy, the joint efficiency of friction stir welding is 76 % and the joint efficiency of gas-metal arc welding is 53 % (Shrivastava, Krones & Pfefferkorn, 2015, p. 163).

5.4.3 Situation Analysis

The situation analysis comprises creating the qualitative representation of the planning task. Hence, object systems, their influential parameters, project characteristics, and relevant actors are specified.

Object System

The model of the object system contains the hierarchical and functional structure of the technical system. The case considers the manufacturing of metal products and, therefore, belongs to the industrial sector "fabricated metal production". The manufacturing is organized in a job shop. Regarding the technical objects, the focus is on welding processes, while media supply and disposal (*e.g.*, air ventilation system) are not analyzed. Interdependencies to pre- and post-processes are considered. Hence, the hierarchical level of the case study is the work center and the segment, *i.e.*, the combination of several processes to a process chain. The primarily addressed function is the manufacturing system with a focus on joining.

The primary energy function of welding equipment regarding the electricity is consumptive usage, *i.e.*, electricity is used to generate the welding heat. However, welding equipment may need to transfer the electricity from alternating current into direct current. Therefore, energy conversion is the secondary energy function.

Creating the functional model of welding equipment includes defining input and output energy flows as well as the influential parameters on energy efficiency. In general, input for welding processes contains electricity, the consumable or non-consumable electrode and shielding gas, whereof electricity is focused in this case study (Hutchins, Robinson & Dornfeld, 2013, p. 538). The output, which may contain radiation, heat, fumes, gases, and waste, is not further considered in this case.

Energy Efficiency Influential Parameters

The energy consumption of welding equipment may be described by power level and time spent in various operational states and may be influenced by welding process parameters. The relevant operational states are operating, idle, and stand-by. The operating state is active while the joint is created, whereas idle contains additional necessary operations (*e.g.*, tool movement). The stand-by state refers to the period of time, when the equipment is switched on without performing an action. Factors that influence the joining energy consumption include process technology, process parameters (*e.g.*, welding speed), auxiliary systems, amount of non-productive time and associated energy demand (*i.e.*, peripheral processes such as shielding gas supply), non-productive time of welding equipment, and load of the equipment (Mose & Weinert, 2015, p. 49). Additional to the welding equipment itself, pre- and post-processes are considered with their total energy consumption. The functional model for the case study is depicted in Figure 38.

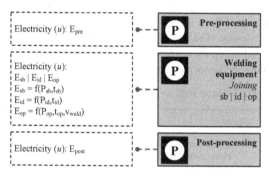

Figure 38: Functional model for case study 1 – planning of welding processes

Project Characteristics

The next step is to determine the characteristics of the planning task. This means to assign the project to the factory planning phases as well as to life cycles of area, building, equipment, and product. The selection and design of a welding process is part of the step determination of functions in the concept planning phase (Krones et al., 2014, p. 232). The functions of the equipment are specified in the detailed planning phase.

Area and building life cycle are not addressed by this case study. For the life cycle of the equipment, the phases concept, development, manufacturing, installation, and operation are relevant. The pertinent phases of the product life cycle are design, work preparation, production planning, and manufacturing.

Actors

The final step of the situation analysis is to identify the actors, their tasks and influences on the energy efficiency influential parameters. Although the case study addresses the planning of welding processes, the later operation is supposed to be prepared. Hence, the actor model contains both planning participants and persons that are affected from planning, *i.e.*, the workers in the welding shop.

The design of the welded part is performed by the product designer, who analyzes the requirements, selects the material and defines the functional structure of the product. During the planning phase, the two main actors are the process engineer and the production engineer. The process engineer determines and arranges the necessary manufacturing steps based on the product design. Furthermore, this role determines details on the separate manufacturing processes. The tasks of the production engineer are focused on the manufacturing equipment including the specification based on the process requirements.

The staff working in the welding shop may be distinguished according to their hierarchical level. The machine supervisor is responsible for the shop and, thereby, supervises the manufacturing processes. The machine setter sets up and programs the machine to manufacture a part. Finally, the machine operator runs the manufacturing equipment.[20] Table 16 summarizes the actors and their tasks.

Table 16: Definition of actors in case study 1 – planning of welding processes

Actor	Tasks
Product designer	– Describes customer requirements – Selects materials – Designs the product
Process engineer	– Selects and conceives manufacturing process – Defines type and order of manufacturing steps – Specifies details of manufacturing processes
Production engineer	– Selects and conceives manufacturing equipment – Prepares specification sheet for equipment manufacturer
Machine supervisor	– Supervises manufacturing processes – Conducts less complex maintenance and repair tasks
Machine setter	– Sets up machine (*e.g.*, tool changes) – Specifies manufacturing parameters
Machine operator	– Puts parts into machine – Operates the manufacturing process – Performs visual checks on product quality

In the following paragraphs, the influence of the roles on the relevant parameters is analyzed. The influence of an actor may vary between the different planning phases or life cycle phases. The product designer is active in the product life cycle phase design and may influence every parameter but the welding process parameters. The process

---

[20] Depending on the degree of automation, the roles machine setter and machine operator may not differ.

engineer determines the manufacturing process, which is relevant in the factory planning phase structure planning and in the work preparation of the product life cycle. Due to the variety of tasks, every parameter may be influenced by this role. The production engineer defines the specification of the equipment, resulting in influences during the factory planning phase detailed planning, and during the concept and development of the equipment. This role may influence every parameter besides the welding process parameters (*e.g.*, welding speed).

The roles machine supervisor, machine setter, and machine operator influence the energy consumption during equipment operation. The machine supervisor is responsible for the workers and them running the machines. Hence, this role may influence the time spent in each operational state. Furthermore, the energy consumption of pre- and post-processes may be influenced since the role is responsible for the entire job shop.

The machine setter adjusts the parameters of the manufacturing process for different jobs and sets up the equipment. Hence, this role effects the welding process parameters and the power level of the operational states. The machine operator influences the time spent in each operational state. Furthermore, the result of the welding process (*e.g.*, in terms of quality) affects the energy consumption of post-processing. Table 17 formally depicts the influences of the product designer on the energy efficiency influential parameters. The influences of the other actors are part of Appendix A1.

Table 17: Assignment of influences on parameters for the role "product designer" in case study 1 – planning of welding processes

		Pre-processing	Welding equipment	Post-processing
Product life cycle phases	Product design	E_{pre}	E_{op}, P_{op}, t_{op}, P_{id}, t_{id}, P_{sb}, t_{sb}	E_{post}
	Work preparation	-	-	-
	Production planning	-	-	-
	Product manufacturing	-	-	-

5.4.4 Identification of Energy Efficiency Measures

Based on the description of the planning task, EEMs are identified using the EEMA algorithm (see Section 4.9). This is done by converting the characteristics of the planning situation into formal vectors and accessing the prototypical knowledge base of EEMs. Having specified each description vector and the influence matrix as part of the actor model, the matching algorithm identifies 22 suitable measures for this planning task (Table 18).

Table 18: Resulting energy efficiency measures in case study 1 – planning of welding processes

Measure	Suitability degree	Product designer	Process engineer	Production engineer	Machine supervisor	Machine setter	Machine operator
Increase welding speed	84 %	–	X	–	–	X	–
Use efficient welding processes	81 %	–	X	X	–	–	–
Reduce number of welding passes	81 %	X	X	–	–	X	–
Select welding process with low requirements for pre-processing	62 %	–	X	–	–	–	–
Use mechanical components with low friction in machinery	60 %	–	–	X	–	–	–
Select welding process with low requirements for post-processing	60 %	–	X	–	–	–	–
Reduce moved masses in machinery	60 %	–	–	X	–	–	–
Design machinery as being able to be switched off and on again	58 %	–	–	X	–	–	–
Reduce welding spatters	57 %	–	X	–	X	X	X
Increase equipment efficiency	57 %	–	–	X	–	–	–
Use inverter-based power supply for welding equipment	56 %	–	–	X	–	–	–
Reduce wire-feed speed	56 %	–	–	–	–	X	–
Equip machinery components with energy-saving mode	55 %	–	–	X	–	–	–
Recover braking energy of machinery	53 %	–	–	X	–	–	–
Decrease welding zone	53 %	X	X	–	–	–	–
Reduce welding temperature	31 %	–	X	–	–	–	–
Use welding processes with high joint efficiency	31 %	–	X	–	–	–	–
Design energy-efficient welding process chains	30 %	–	X	–	–	–	–
Apply energy-optimized motion profile for machinery and equipment	28 %	–	X	X	–	–	–
Switch off energy-intensive machinery in breaks	26 %	–	–	–	X	–	X
Switch off process-dependent components appropriately	25 %	–	–	–	X	–	X
Select proper material for welding processes	23 %	X	–	–	–	–	–

The measures are explained in the following paragraphs.[21] The EEM "Increase welding speed" means to raise the velocity, with which the welding tool is moved along the defined path. The energy consumption to conduct a weld is calculated as (Liu, Li & Shi, 2014, p. 543):

$$E_{weld} = \frac{\ell}{v} \cdot \frac{P_{weld}}{\eta},$$

(5.1)

with ℓ being the weld length, v the welding speed, P_{weld} the welding power and η the joining efficiency, *i.e.*, the ratio between energy entering the workpiece and end energy input (*e.g.*, electricity). A higher welding speed leads to a shorter welding duration. This may decrease energy consumption depending on the effect on welding power. For example, BAHRAMI ET AL. identify a reduced energy consumption through increasing arc power and welding speed (Bahrami, Valentine & Aidun, 2015, p. 119).

The EEM "Use efficient welding processes" considers the energy efficiency of the welding process. Energy efficiency represents the ratio between a useful output and the required energy input. The output of a welding process is represented by a weld of defined length and quality, which requires an energy input into the workpiece. The energy input into the joining process, however, is higher due to energy losses. As such, the equipment efficiency and thermal efficiency are important indicators to describe these losses (Shrivastava, Krones & Pfefferkorn, 2017).

The equipment efficiency accounts for losses through the conversion of energy, such as the transformation of voltage. Hence, it describes the ratio between the generated joining energy and the consumed end energy (Hälsig & Mayr, 2013, p. 288).

The thermal efficiency represents the ratio between heat input into the workpiece and joining energy and, thus, accounts for losses through heat transfer from the source to the workpiece (Shrivastava, Krones & Pfefferkorn, 2017). Reference values for the thermal efficiency of selected processes is provided by the standard DIN EN 1011 (Table 19). However, this standard does not consider the influence of welding process parameters nor the welded material pairing. Furthermore, efficiencies are only described with reference to submerged arc welding (which is normalized by an efficiency of 1).

The welding efficiency η_{weld}, which integrates both equipment and thermal efficiency, may be described as (Hälsig, Kusch & Mayr, 2012, p. 98):

$$\eta_{weld} = \frac{Q_{BM} + Q_{FM}}{P_w \cdot t_w},$$

(5.2)

---

[21] Since the assessment with EEMIS sheets is not part of this case study, some additional information on the measures is given in the text.

whereof Q_{BM} is the heat input into the base material, Q_{FM} the heat input by the filler material, P_w the welding power, and t_w the welding time. Depending on the choice of process parameters, HÄLSIG ET AL. identify joining efficiencies as depicted in Table 20 (Hälsig, Kusch & Mayr, 2012, p. 104).

Table 19: Thermal efficiency of selected welding processes (DIN EN 1011, p. 12)

Welding process	Thermal efficiency
Submerged arc welding	1
Metal inert gas welding	0.8
Metal active gas welding	0.8
Tungsten inert gas welding	0.6
Plasma welding	0.6

Table 20: Efficiencies of selected welding processes (Hälsig, Kusch & Mayr, 2012, p. 104)

Welding process	Minimum efficiency	Maximum efficiency
TIG welding	0.68	0.79
Plasma welding	0.69	0.80
MAG welding (steel electrode)	0.68	0.86
MIG welding (aluminum electrode)	0.73	0.91

Yet, even for one process, there may be a wide range of values for energy intensity, depending for example on material thickness. Hence, the following sources give only a broad overview of joining energy intensities of various welding processes. RIEDEL describes a comparison of various welding processes for manufacturing magnesium car doors (Riedel, 2013, p. 22): The specific energy consumption accounts for 148 kJ/m for MIG welding, 55 kJ/m for laser-hybrid welding, and 20 kJ/m for laser welding. LIU ET AL. compare the energy consumption between gas-tungsten arc welding, laser welding, and laser-GTA hybrid welding for MgAl3Zn alloy. The resulting joining efficiencies are 339 J/m for GTAW, 972 J/m for laser welding and 216 J/m for the hybrid process (Liu, Li & Shi, 2014, p. 543). The authors conclude that a combination of laser and arc welding can increase energy efficiency. SHRIVASTAVA ET AL. analyze parameters that influence the energy consumption of friction stir welding (Shrivastava, Overcash & Pfefferkorn, 2015, p. 51): The specific energy consumption for friction stir welding for AlMg1SiCu is identified 10 to 35 J/mm^3 and 20 to 60 J/mm^3 for AlZn5.5MgCu, whereof the range is caused by different welding feed rates.[22] SPROESSER ET AL. study the environmental effects of MMAW, LAHW, and GMAW when welding thick-metal plates of low alloyed

---

[22] It should be noted that the cross-sections are different for each alloy.

structural steel. The identified energy consumption is 0.9 kWh/m for LAHW, 2.1 kWh/m for GMAW and 3.9 kWh/m for MMAW, respectively (Sproesser et al., 2015, p. 50).

"Reducing the number of welding passes" is a measure to increase energy efficiency since the number of welding passes is directly related to the welding energy consumption. It depends on the geometry of the part, the capability of the technology to create thick welds and the selected welding process parameters. The number of required welding passes is especially influenced by the material deposition rate, *i.e.*, the rate at which the filler material is deposited. For example, SPROESSER ET AL. compare the energy consumption for welding 20 mm thick plates of low alloy steel between a conventional gas-metal arc welding process and one with a modified spray-arc and a reduced flange angle. This modified process increases welding power but only requires half the number of welding passes, such that the energy consumption can be reduced by 38 % (Sproesser et al., 2015, p. 50).

The two measures "Select welding process with low requirements for pre-processing" and "Select welding process with low requirements for post-processing" address the design of energy-efficient process chains, since the welding process itself influences the requirements for pre- and post-processes. Pre-processing may include pre-heating, edge preparation, material control, cleaning, or surface treatment. The effort for these processes varies with the requirements of the welding process. For example, complex manufacturing process chains in the automotive industry (*e.g.*, punching, pressing, forming, welding) lead to an increasing tolerance chain (Trommer, 2010, p. 62). When the welding process has high tolerances towards differences in the gap width, less effort is required for pre-processing. On the contrary, when the welding process is incapable of bridging an existing gap, additional corrections are necessary. Post-processing may include cutting, grinding, straightening, or heat treatment, which again depends on the selected welding process. For example, a low heat input reduces the material stress, which may reduce the need for straightening (Shrivastava, Krones & Pfefferkorn, 2015, p. 160).

A measure with regard to the welding equipment is to "Use mechanical components with low friction in machinery". Friction is, in general, a cause for higher energy consumption of machinery. However, this EEM has only a low relevance for the specific case. A possible application scenario is to reduce friction in robot joints.

Another EEM for the equipment consists of "Reducing moved masses in machinery". Moving masses causes energy consumption, which is important for movable parts in machinery. With regard to welding equipment, this may refer to robot arms of automatic welding machines.

The machine design needs to provide a possibility to switch on and off equipment easily ("Design machinery as being able to be switched off and on again"). This EEM is a general requirement for production systems without a specific focus on welding. The purpose is to switch off equipment during breaks or weekends in order to reduce idle and stand-by energy consumption. For this measure, the production engineer needs to provide the possibility in the machine design and the machine operator and/or supervisor need to pay attention to the realization during equipment operation.

An interdisciplinary relevant measure is to "Reduce welding spatters". Welding spatter are small metal particles from the electrode that are expelled during welding but do not form a part of the weld. Possible causes for spatter are incorrect process parameters (*e.g.*, wire-feed speed, angle of welding torch), contaminated workpiece surfaces (*e.g.*, oil), or quality of consumables, *i.e.*, wire and shielding gas. Possible negative effects of spatter are that they may be deposited in the welding nozzle and, hence, may lead to defects and failures. Furthermore, some application areas require welding products to be free of spatter, *e.g.*, security-related parts. However, mechanically removing spatter is time- and energy-consuming. Therefore, welding spatter should be reduced through process selection, choice of process parameters, and properly operating the welding equipment. An example of a nearly spatter-free welding processes is cold metal transfer welding (Trommer, 2010, p. 62).

Furthermore, it is important to "Increase equipment efficiency". The equipment efficiency is defined as the ratio between the generated joining energy and the consumed energy of welding equipment. It may be increased by an appropriate selection of welding equipment. One measure to increase equipment efficiency is to "Use inverter-based power supply for welding equipment". This technology reduces heat losses and power demand during idle time: While old equipment requires a power of approximately 1 kW, inverter-technology based equipment needs as low as 50 W during idle mode (Lehnertz, 2009, p. 28).

"Reducing wire-feed speed" means to lower the velocity to feed the consumable electrode of a fusion welding process. The proper wire-feed speed needs to be selected depending on the material and weld geometry. An increase of wire-feed speed leads to a higher welding current, which transfers more heat into the workpiece. This increases heat losses, leading to a reduced efficiency of the welding process. (Hälsig, Mayr & Kusch, 2016, p. 263)

The EEM "Equip machinery components with energy-saving mode" is a general measure that may be applied for any production system. Since equipment is not operating continuously, it should be equipped with an energy-saving mode that has a low power demand and may be activated during process breaks. For example, an analysis in an automotive car body shop identifies that welding robots spend only 26 % time moving. Therefore, a robot manufacturer equips the robots with three different stand-by operational states in order to adjust the state of the components (*e.g.*, control, brakes) to the duration of a movement break. This reduces the stand-by energy consumption by up to 80 %. (Senft, 2012, p. 3)

"Recovering braking energy" is a general principle to reduce energy consumption of processes, in which masses are moved. In welding processes, this may be applied for welding robots (Senft, 2012, p. 2): Since the drives of a robot are performing functions in different operational states, energy may be recovered from one braking axis to other accelerating axes in the robot system.

A measure involving the product designer is to "Decrease the welding zone". The welding zone is understood as the cross-sectional area of the weld and is affected by both the part geometry and the welding technology. In general, a smaller weld area reduces the mass of material that needs to be melted in fusion welding processes. Hence, the required thermal energy is reduced. Furthermore, welding process technologies differ with regard to their energy density, *i.e.*, high energy density welding processes focus the energy to a small welding area, which results in a low heat input into the workpiece. Laser beam welding and electron beam welding are categorized as high energy density welding processes (Phillips, 2016, p. 114). The efficiency of laser welding is low due to high losses when converting net energy to processing power. However, it allows to apply higher welding speeds, which reduces lead times. Furthermore, the localized heat input reduces the effort for post-processing (Kaierle, Dahmen & Güdükkurt, 2011).

"Reducing the welding temperature" decreases the heat input into the workpiece, hence, the energy that is required to produce the heat. As such, solid state welding processes are considered to be more energy-efficient than fusion welding methods. For example, friction stir welding may save 30 % of the energy consumption as compared to gas-metal arc welding (Shrivastava, Krones & Pfefferkorn, 2015, p. 165).[23] Moreover, lower welding temperatures lead to different mechanical and microstructural properties of the welded material. For example, friction stir welding creates a fine microstructure with good mechanical properties due to the plastic deformation below the melting point of the material (Mishra & Ma, 2005, p. 2).

"Using welding processes with high joint efficiency" addresses the process selection. The joint efficiency is an indicator to assess the weld quality (see Section 5.4.2). A high tensile strength in the as-welded part allows to make a workpiece thinner in order to withstand the same maximum tensile force. This reduces the required material input. Furthermore, a lower welding depth reduces the energy consumption for welding since it may need a smaller number of welding passes and may reduce the effort for pre-processing (*e.g.*, edge preparation). Finally, lower product weight reduces the energy consumption of transport and handling processes.

The "Design of energy-efficient process chains" means to combine manufacturing processes in one step or to integrate peripheral processes into the manufacturing process (Müller et al., 2009, p. 126). Process chains may be compared by assessing the total

---

[23] It should be noted that this saving effect does not only result from the lower welding temperature.

energy input to manufacture a specific part. An example with regard to welding is to reduce the energy consumption for manufacturing a workpiece with laser cutting and welding by 90 % as compared to milling and drilling (Kettner-Reich, 2011, p. 28). Another example for a successful process chain substitution is described by an automotive supplier company, which produces sheet steel rings by forming and welding instead of punching and achieves 73 % material savings (Itasse, 2014, pp. 20 f.).

An equipment-related measure is the EEM "Apply energy-optimized motion profile for machinery and equipment". This is both a specification for the equipment manufacturer and an instruction for the process engineer. Programming the movement trajectory of a robot is usually adjusted to the shortest path, whereas the most energy-efficient trajectory may vary considerably (Senft, 2012, p. 2).

An EEM during the operation of welding equipment includes to "Switch off energy-intensive machinery in breaks". This relates to the measure to design machinery that can be switched off. A similar measure is to "Switch off process-dependent components appropriately". It may generally be applied to increase energy efficiency of production equipment and means that the peripheral components (e.g., cooling, ventilation) should only be switched on when manufacturing equipment is operating.

Finally, the product designer influences the energy consumption through the selection of material ("Select proper material for welding processes"). The energy that is required to melt a material is determined by its heat capacity and melting enthalpy. For example, aluminum has a lower melting point than steel.

5.4.5 Interpretation of Case Study Results

The case study demonstrates the identification of energy efficiency measures for a planning task on manufacturing processes based on the qualitative analysis including the object system, its energy efficiency influential parameters, and the influence of various actors. The object system contains the welding equipment and considerations on relevant pre- and post-processes. The parameters to describe the energy consumption are, in general, power levels and time spent in various operational states (e.g., operation, idle, stand-by). Furthermore, welding process parameters, such as welding speed, influence the energy efficiency.

The actors considered are the product designer, process engineer, production engineer, machine supervisor, machine setter, and machine operator. The influence of these actors on the energy consumption is modeled with regard to the various planning phases and life cycle phases of both product and equipment. The actors contain both planning participants and persons that are involved in the equipment operation.

Based on the qualitative analysis, 22 measures are identified as being relevant for the case study. These measures are assigned to the actors, whereof the process engineer and the production engineer are the most important actors with 10 and 9 measures, respectively. This emphasizes the relevance of considering energy efficiency during the planning of production processes and equipment.

Approximately 40 % of the measures are assigned to more than one actor, which emphasizes the need of interdisciplinary cooperation. This results from the fact that the energy consumption of a system is influenced by the complex interrelationship between various parameters, such as the efficiency of a welding process, which is determined by the welding technology, equipment, and concrete choice of process parameters. Furthermore, measures need to be planned ahead and applied appropriately during operation, which underlines the need of cooperation between planning participants and operative staff. For example, the number of required welding passes is in general determined by the specifications of the welding process (process engineer), but in detail influenced by the concrete program control (machine setter).

The measures are assessed with the suitability degree that expresses the applicability in the planning task. In this case, the measures relate to the factory planning phases as well as to product and equipment life cycle phases. Therefore, the suitability degree is composed of the evaluations for each of these components. Hence, a measure is assessed with a higher value when it addresses several components.

For example, the EEM "Reduce number of welding passes" addresses the three components, *i.e.*, the factory planning phase structure planning, the design phase of the product life cycle, and the operation phase of the equipment life cycle. On the contrary, the EEM "Select proper material for welding processes" only relates to the product design, which leads to a lower suitability degree. Thus, the suitability degree shows a tendency on more relevant and less relevant measures.

As a result, the methodical approach helps to structure the influence of various actors based on a rough task definition and supports the identification of appropriate energy efficiency measures that may be applied during planning or later operation.

5.5 Case Study 2: Planning of Logistics Systems

The second case study addresses the energy efficiency of logistics systems.[24] Following the procedure model in Section 4.10, the steps goal definition, situation analysis, measure identification, and measure assessment are performed.

---

[24] The case study is based on an industrial project between the Chemnitz University of Technology and the AUDI AG, which was conducted between September 2015 and January 2016. Parts of the case have been published in Krones & Müller, 2016; Krones et al., 2016; Hopf et al., 2016.

5.5.1 Goal Definition

The purpose of the case study is to identify EEMs in planning and operating a logistics building for order picking in the automotive industry. The order picking connects the warehouse with the automotive assembly. Its aim is to provide logistic units (*i.e.*, assembly parts) according to the desired amount and sequence as defined by the production process. Currently, the order picking is most commonly realized with manual processes within so-called supermarkets. An automotive supermarket is a logistics system that is placed close to the manufacturing area and that handles assembly material (Klug, 2010, p. 197). The manual order picking process is based on the "person-to-goods" strategy, *i.e.*, the employee moves to the goods in the warehouse area. Main advantages of this strategy are low investment effort, flexibility towards fluctuating requirements in the amount of material, and the possibility to handle a variety of parts (Klug, 2010, p. 193). However, the productivity is quite low since the employees spend a significant share of their work time walking.

On the contrary, the "goods-to-person" strategy means to transport the goods to an employee who operates on a stationary workplace. An advantage of the concentration of workplaces lies in the opportunities to support the order picking, for example through pick-by-light technologies. Furthermore, the static structure of rack positions is relaxed, which makes the warehouse easier to adapt to changes in amount and variety of assembly parts. The "goods-to-person" concept allows to automate portions of the order picking process, *e.g.*, the retrieval of parts. This reduces the width of aisles and the required floor space. Both concepts for order picking are visualized in Figure 39.

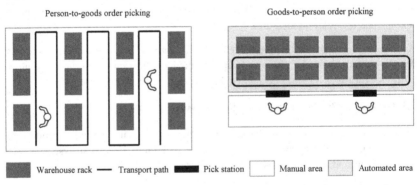

Figure 39: Comparison between person-to-goods order picking concept and goods-to-person order picking concept (adapted from Krones et al., 2016, p. 20)

The case study is supposed to identify EEMs for planning a logistics system, in which the order picking is realized with a "goods-to-person" strategy. The building consists of two areas: The pick area contains the workplaces of the employees. Each workplace is either used to pick material prior to transporting it to the assembly area or to load material into the warehouse racks. The second area contains the warehouse racks.

Within the warehouse area, automated guided vehicles (AGVs) transport the racks that are filled with material to the pick stations. Furthermore, empty racks are transported to the loading stations. The task of the planning project is to layout the building, to plan the workplaces, and to develop specifications for the equipment.

The energy efficiency of a logistics system is determined by the ratio between a logistical performance and the required energy input. The logistical performance may be characterized by, among others, the number of transported pieces, the transported weight, and/or the distance traveled. In this case study, the energy consumption is focused. Hence, the purpose is to identify measures that reduce the energy consumption, whereas the effects on the logistical performance are not considered in detail (Krones, Hopf & Müller, 2014, p. 501). Reducing the energy consumption of a logistics system may be achieved by changes at the technology or through an optimized logistics operation (Müller et al., 2013b, pp. 153 f.).

5.5.2 Situation Analysis

The situation analysis means to qualitatively describe the domains object system, energy efficiency influential parameters, project characteristics, and actor.

Object System

At first, the object system is modeled, which includes the general factory information as well as the hierarchical and functional structure of the technical system elements. The case study is carried out in a large enterprise of the automotive industry. The production of automobiles is realized as a series production in a flow shop.

According to the hierarchical model, the case is assigned to the level of a building. The building consists of the divisions production system (*i.e.*, order picking system), building system, building services, and process technology. Within the production system, the relevant work centers are the AGVs, which are summarized as AGV fleet on the level of a segment. The building system is represented by the building structure (*e.g.*, walls, doors). The building services include heating, lighting, and information and communication equipment.

Process technology comprises the systems that supply the production system, *i.e.*, the battery charging system that provides the electrical energy for the AGV, and the safety equipment. Figure 40 presents the considered systems in their hierarchical order.

For the functional description, the systems' functions with regard to material and energy flows are analyzed. This includes defining the input and output flows and distinguishing the systems' operational states.

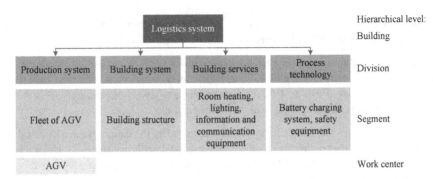

Figure 40: Hierarchical system structure for case study 2 – planning of logistics systems

The logistics system stores material for the automotive assembly (main function material storage). The required energy carriers are electricity and heating gas. The entire building may be operating (during the work time), in stand-by state (*e.g.*, during breaks), or non-operating (*e.g.*, during weekends or production-free periods). The systems on the hierarchical level of the division are not further explained since these do not add information to the level of segments and work centers.

The hierarchical level segment consists of the systems AGV fleet, building structure, heating, lighting, information and communication equipment, battery charging system, and safety equipment. The AGV fleet performs the material transport by using electricity. The operational states of the AGV fleet refer to the states of the entire logistics system. The building system is supplied by the building services with the purpose of maintaining proper work conditions (*e.g.*, room climate). Hence, the specifications of the building structure influence the energy consumption of room heating and lighting. The room heating consumes heating gas in order to provide the output flow of heat. A single operational state is assumed as an average to facilitate matters, since the differentiation is based on the weather and, hence, cannot be influenced in the case.

The lighting consumes electricity and may either be operating or non-operating. The information and communication equipment is separated into the states of operating and non-operating.[25] The battery charging system has the main function of energy conversion with electricity as both input and output flow. It shows two operational states, *i.e.*, charging and stand-by. The safety equipment is operated continuously, *i.e.*, has only one operational state.

On the hierarchical level of the work center, the only considered system is the AGV. Its main function of material transport is realized while consuming electricity. The

---

[25] The information and communication equipment includes both servers, which are operated continuously, and workplace computers, which are operated depending on the users.

operational states are operating (*i.e.*, driving or waiting during the transport process), stand-by (*i.e.*, during breaks), and non-operating (*e.g.*, during weekends).

Energy Efficiency Influential Parameters

The next step is to identify the energy efficiency influential parameters for each system. Main influences on the power load of an AGV are transport velocity, acceleration, mass of the goods, mass of the AGV, and the rolling resistance coefficient (Krones & Müller, 2016, p. 1097). Furthermore, the energy consumption for a transport cycle is influenced by traveling distance d_T, proportion of empty trips α_{emp}, and number of acceleration processes n_A (Müller, Hopf & Krones, 2013, p. 54). The energy consumption in each operational state is represented by the following equations (adapted from Müller, Krones & Hopf, 2012, p. 369; Krones & Müller, 2016, p. 1098):

$$
\begin{aligned}
E_{no}^{AGV} &= P_{no} \cdot t_{no} \\
E_{sb}^{AGV} &= P_{sb} \cdot t_{sb} \\
E_{op}^{AGV} &= P_{sb} \cdot t_{op} + 0.5 \cdot (m_{AGV} + m_G) \cdot v_T^2 \cdot n_A + C_{rr} \cdot (m_{AGV} + m_G) \cdot g \cdot d_T.
\end{aligned}
\tag{5.3}
$$

The energy consumption both during non-operating mode and stand-by mode is composed of the power level (P_{no} or P_{sb}) and the time spent in this operational state (t_{no} and t_{sb}). The energy consumption while operating E_{op} is composed of three components: First, the stand-by components are active during driving time t_{op} and require the power P_{sb}. The second component refers to the energy to overcome the acceleration resistance, which is influenced by the mass of the vehicle m_{AGV}, the mass of the goods m_G, driving velocity v_T, and the number of acceleration processes n_A during the cycle. The third component accounts for the rolling resistance and is effected by the rolling resistance coefficient C_{rr}, the masses m_{AGV} and m_G, and the driving distance d_T.

The relation between the system's function, its operational states, and energy efficiency influential parameters is represented in Figure 41. It shows the input of using electricity to operate the AGV. The main function of the system is material transport, which is indicated by the letter "T" in the box. The operational states are non-operating (no), stand-by (sb), and operating (op). On the input side, the influential parameters are visualized for the entire system and for each operational state.

The functional models for the AGV fleet, battery charging system, building system, room heating, lighting, information and communication equipment, and safety equipment are enclosed in Appendix A2.

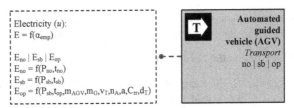

Figure 41: Functional model of the AGV for case study 2 – planning of logistics systems

Project Characteristics

The next step is to determine the characteristics of both the factory planning project and the energy efficiency project. Due to the generality of the study, the requirements of the energy efficiency project are not limited in terms of budget and/or implementation time. The factory planning project is characterized by an early concept stage, $i.e.$, the concept planning phase. However, since the case study focuses on a holistic perspective on the goods-to-person order picking, aspects of other planning and management tasks are considered as well. Hence, the parameters to represent the project characteristics are determined as structure planning, dimensioning, ideal planning, real planning, detailed planning, operation, and maintenance.

Actors

The final step of the situation analysis is to create the actor model. This includes the definition of actor-specific tasks and the effect of each actor on the influential parameters. The main actor that develops the concept is the logistics planning. This department closely works together with the logistics management, whose task it is to operate the order picking system ($i.e.$, responsible for a warehouse area that delivers parts to the automotive assembly). The definition of requirements for the transport process is a task that requires both the logistics planning and the AGV manufacturer. The order picking staff are the persons directly working with the system. The maintenance staff is responsible for ensuring proper working of the technical system. Building planners are responsible for the building's structural elements ($e.g.$, walls, floors). Finally, the infrastructure including heating and lighting is planned by the building services staff. Table 21 summarizes the actors and their tasks in this case.

Having defined the actors, their effect on the energy efficiency influential parameters throughout the planning and life cycle phases is analyzed. As introduced above, the relevant life cycle phases are the planning and operation of both building and equipment. More specifically, the planning phases comprise the concept planning (structure planning, dimensioning, ideal planning, real planning), detailed planning, operation, and maintenance. An actor's influence is represented by a matrix that indicates the possibility to modify a parameter in a specific phase.

Table 22 shows this influence matrix exemplary for the role of logistics planning. The other role representations are contained in Appendix A2.

Table 21: Definition of actors in case study 2 – planning of logistics systems

Actor	Tasks
Logistics planning	– Specifies logistical requirements (*e.g.*, transport intensities) – Defines logistics concept (including racks and containers) – Selects logistics resources – Specifies required building layout – Determines working requirements (*e.g.*, ambient conditions)
Logistics management	– Initiates and controls transport orders
Order picking staff	– Performs order picking tasks at the pick station
Maintenance staff	– Monitors the technical condition of equipment – Performs service and repair
Building planning	– Designs the building system (*e.g.*, walls, windows, floors)
Building services staff	– Handles technical building services (*i.e.*, heating and lighting)
AGV manufacturer	– Determines a vehicle's specifications – Selects battery technology and manufacturer – Provides software for logistics fleet management

Table 22: Assignment of influences on parameters for the role "logistics planning" in case study 2 – planning of logistics systems

Object system	Concept planning	Detailed planning
AGV fleet	n_{AGV}, P_{op}	t_{no}, t_{sb}, P_{op}
AGV	t_{dr}, d_T	d_T, m_G, v_T, α_{emp}, n_A, t_{no}, t_{sb}
Battery charging system	-	E_{ch}
Room heating	θ_i, A, V, t_{ae}	-
Lighting	E_m	-
Information and communication equipment	-	P_{op}, P_{no}, t_{op}, t_{no}
Safety equipment	-	P_{op}, P_{no}, t_{op}, t_{no}

The main task of logistics planning is to define the logistics concept and to select the logistics resources, which includes the AGV and the warehouse racks. Hence, the main influence is on the energy consumption of the AGV. The number of vehicles is determined by the logistics planning based on the number of picks per hour and the average number of picks for each warehouse rack. The layout planning and the spatial arrangement of the warehouse racks influences the transport distance. The mass to be transported is influenced, for example, by the design of the warehouse racks and the choice of containers. The control of the AGV fleet during operation affects the share of time spent in the various operational states and the proportion of empty trips.

Besides, the logistics planning determines the room area and, hence, influences the energy consumption of the building services, i.e., heating and lighting. Furthermore, the conditions of the work process set requirements for the building services, such as room temperature and illumination. Finally, the number of workplaces, i.e., pick stations and load stations, defines the required equipment for information, communication, and safety.

5.5.3 Identification of Energy Efficiency Measures

The next step is to assign EEMs by using the algorithm described in Section 4.9. The application field of the case study is subdivided into the following areas: The main part is the logistics process using the "goods-to-person" order picking concept. Hence, the holistic consideration of other planning and management phases only comprises the logistics task. Furthermore, the building is the second analyzed field. For this object area, only the early concept planning needs to be considered. Thus, the formal assignment task is performed in two steps, which distinguish between the logistics process and the logistics building.

The object system and project characteristics for the area of the logistics process are defined as follows: The technical components comprise the divisions production system and process technology. Hence, the hierarchical level contains the division, the segment (AGV fleet, battery charging system), and the work center (AGV). The systems' functions are material transport, information and communication as well as conversion and consumption of electricity. The considered phases are structure planning, dimensioning, ideal planning, real planning, detailed planning, operation, and maintenance.

Taking this information, the input vectors for the technical object system and the project characteristics are generated. The measure assessment criteria are not relevant to describe the planning situation due to the generality of the task. The actor model is formulated according to the influences of the roles on the parameters as identified during the situation analysis. The building services staff is not considered for the logistics process, since this role has only tasks regarding the logistics building.

The assignment task for the logistics process results in 17 measures, which are depicted in Table 23. It contains the measures, their suitability degree and the assignment to the respective actors. For better understanding, the number of the measure is marked with a category ("T" for transport). The measures address the AGV, the warehouse, the battery charging system, and the information and communication equipment.

Table 23: Resulting energy efficiency measures in case study 2 – planning of logistics systems for the area of the logistics process

No.	Measure	Suitability degree	Logistics planning	Logistics management	Order picking staff	Maintenance	Building planning	AGV manufacturer
T2	Reduce masses to be transported	68 %	X	X	–	–	–	–
T9	Increase utilization rate of transport containers	60 %	X	X	–	–	–	X
T18	Reduce acceleration processes	53 %	X	X	–	–	–	X
T8	Use vehicle fleet management	51 %	X	X	–	–	–	X
T23	Reduce driving velocity	49 %	X	–	–	–	–	–
T16	Split floor conveyors into low and high performance vehicles	45 %	X	–	–	–	–	–
T24	Arrange warehouse racks in order to minimize transport routes	39 %	X	–	–	–	–	–
T21	Summarize transport orders	34 %	–	X	–	–	–	–
T22	Reduce unnecessary movements of logistics vehicles	29 %	–	X	–	–	–	–
T127	Regularly carry out a compensation charge of batteries	29 %	–	–	–	X	–	–
T6	Increase ratio between load capacity and total mass	29 %	X	–	–	–	–	X
T20	Switch equipment off or into stand-by mode appropriately	28 %	–	X	X	–	–	–
T25	Reduce areas for transport and providing material	28 %	X	X	–	–	–	–
T19	Reduce mass of containers and packaging material	24 %	X	–	–	–	–	–
T11	Use Ri charging process for lead acid batteries	23 %	–	–	–	–	–	X
T13	Use high frequency charging equipment with electrolyte circulation	23 %	–	–	–	–	X	X
T15	Reduce rolling resistance of logistics vehicles	22 %	–	–	–	–	–	X

The main parameters to influence the energy consumption of an AGV concern the mass of transported goods, rolling resistance, acceleration, and velocity. Masses are influenced from several directions: First, this relates to the mass of the transported goods, which may be influenced by logistics planning and management ("Reduce masses to be transported"). Secondly, the ratio between the load capacity and mass of a vehicle is important, which is influenced by the AGV manufacturer ("Increase ratio between load capacity and total mass"). Finally, the mass of containers and packaging material needs to be transported ("Reduce mass of containers and packaging material"). In this case study, the warehouse racks are lifted and transported by the AGV. Hence, the mass of the warehouse racks influences the energy consumption as well.

On the contrary, an EEM is suggested to "Increase the utilization rate of transport containers". Although this increases the transport masses, space requirements may decrease, which reduces the energy consumption of building services.

Reducing the energy consumption of the AGV may be achieved by "Reducing the acceleration processes". The acceleration is adjusted by the manufacturer, whereas the frequency of acceleration processes is influenced by logistics planning (*e.g.*, one-way movement) and logistics management (*e.g.*, waiting queue at the pick stations). Moreover, energy demand decreases when "Reducing the driving velocity". This needs to be determined in cooperation between logistics planning and AGV manufacturer.

Considering the operational tasks of the AGV, it is suggested to "Use a vehicle fleet management", which assigns transport tasks to vehicles with regard to economic objectives. Additionally, the dimensioning of the fleet is important. For example, varying transport tasks may ask for a heterogeneous fleet ("Split floor conveyors into low and high performance vehicles").

The transport distance greatly influences the energy consumption of an AGV. Hence, it is helpful to "Arrange warehouse racks in order to minimize transport routes" as a task of layout planning. During operation, the transport distance is influenced by "Summarizing transport orders". This has the additional effect of increasing the vehicles' utilization rate. Logistics management needs to "Reduce unnecessary movements of logistics vehicles", such as shunt movements.

Both logistics and information and communication equipment should be switched off when not used for the process ("Switch equipment off or into stand-by mode appropriately"). This measure needs to be pursued by logistics management and order picking staff.

The main lever to reduce energy consumption in the warehouse area is the required space ("Reduce areas for transport and providing material"). Hence, areas should be reduced in cooperation between logistics planning and building planning. Eventually, the layout planning, *i.e.*, arranging the warehouse racks, influences the transport needs.

The battery charging system is provided by the AGV manufacturer. Proper maintenance of the battery charging system needs to be conducted by the maintenance staff through "Regularly carrying out a compensation charge of batteries". This accounts for differences of the charging status between the battery cells. The measures "Use Ri charging process for lead acid batteries" and "Use high frequency charging equipment with electrolyte circulation" address the technical specifications of the charging process in order to make it more efficient.

Finally, it is suggested to "Reduce the rolling resistance of logistics vehicles". The rolling resistance results from the material and geometry of the wheels (AGV manufacturer) and the flatness of the floor (building planning).

The second field for assigning EEMs relates to the logistics building. The technical object system is described by the divisions building services and building structure. The building services contain the segments heating and lighting. The energy function of the system is the consumption of electricity (lighting) and gas (heating). The planning phases contain structure planning, dimensioning, ideal planning, and real planning. Furthermore, the construction phase of the building life cycle is regarded since the case study addresses the planning of a new building. The actor model is principally the same as for the logistics process, but only the roles logistics planning, building planning, and building services staff are relevant for this part.

The assignment task for the logistics building results in 12 measures, which are depicted in Table 24 and explained briefly in the following paragraphs. The table contains the measures, their suitability degree, and the assignment of actors. The number of the measure is complemented with a category ("T" for transport, "L" for lighting, and "H" for heating). The measures "Reduce areas for transport and providing material" and "Arrange warehouse racks in order to minimize transport routes" are mentioned above with regard to the logistics process. However, these are also results for the logistics building since the area might be influenced, which effects the energy consumption for heating and lighting. Furthermore, the transport processes into and outside of the building influence ventilation heat losses and, thereby, increase the heat energy demand of the building.

The lighting energy consumption is influenced by daylight use and illumination. Hence, it is important to adjust the lighting to the requirements of the work process, which may include reducing illumination or varying between the general, relatively dark, general lighting for the warehouse area, and a more intensive workspace lighting ("Split lighting into general and workspace lighting" and "Adjust illumination to work task"). Furthermore, planning a building needs to consider the usage of daylight during later operation ("Place windows in order to maximize daylight use").

Table 24: Resulting energy efficiency measures in case study 2 – planning of logistics systems for the area of the logistics building

No.	Measure	Suitability degree	Logistics planning	Building planning	Building services
T25	Reduce areas for transport and providing material	53 %	X	X	–
H12	Reduce heat losses through in-bound and out-bound transport processes	47 %	X	X	–
L27	Split lighting into general and workspace lighting	46 %	X	–	X
H30	Use decentralized heating systems	46 %	–	–	X
L26	Adjust illumination to working task	42 %	X	–	X
L125	Place windows in order to maximize daylight use	42 %	–	–	–
H81	Use radiating heating systems	40 %	–	X	X
H92	Reduce average room temperature	39 %	X	–	–
H169	Reduce ratio between surface and volume of a building	35 %	–	X	–
L73	Reduce height of lamps	32 %	–	–	X
H29	Reduce room height	28 %	X	X	–
T24	Arrange warehouse racks in order to minimize transport routes	28 %	X	–	–

Finally, lighting is necessary on the level of the workstations, which means that it is more efficient to focus illumination on these points, *i.e.*, to "Reduce the height of the lamps".

The heating energy consumption is influenced by the room temperature, the opening cross sections (*e.g.*, for gates), the room height, and transmission heat losses ("Reduce heat losses through in-bound and out-bound transport processes", "Reduce average room temperature", "Reduce room height" and "Reduce ratio between surface and volume of a building"). The technology for distributing the heat into the room needs to be selected properly (for example through "Using radiating heating systems"). Since the pick stations take up only parts of the warehouse area, the "Use of decentralized heating systems" may be a relevant alternative.

5.5.4 Assessment and Selection of Energy Efficiency Measures

The result of the algorithmic assignment is a list of 27 energy efficiency measures.[26] The next step is a review of these measures for the specific case in order to qualitatively evaluate their usefulness. The input information is provided by the measure implementation sheets and additional industrial application examples.

Structure and content of the measure implementation sheets is illustrated exemplary for the EEM "Reduce heat losses through in-bound and out-bound transport processes" (Figure 42). The energy efficiency measure implementation sheets for the other measures are depicted in Appendix A2.

The measure addresses heat losses and, hence, is assigned to the object area heating. Heat losses may have two causes in general, which is explained by the theoretical background of the sheet: Transmission heat losses as the first category are caused by heat conduction through closed surfaces. Secondly, ventilation heat losses result from heat convection (*e.g.*, air draft). Gates and doors in logistics buildings lead to two types of ventilation losses: The first type occurs while they are closed, *i.e.*, due to leakage, whereas the second type is caused by opening a gate (*e.g.*, during material delivery).

Measures to reduce either transmission or ventilation heat losses include: reduce heat transition coefficient and leakage class, reduce opening time of gates, use air locks, use high-speed doors, adjust opening height, and insulate gates during delivery. Benefits in terms of energy savings through applying the measures are given by a scientific study.

Furthermore, an industrial application example on this measure is given (Figure 43). It shows the installation of gates with a low heat transition coefficient at the logistics center of a company that produces agricultural vehicles.

---

[26] Seventeen and twelve measures are identified in the area of the logistics process and the logistics building, respectively. Two of the measures may be assigned to both categories.

Energy efficiency measure					
Reduce heat losses through in-bound and out-bound transport processes					

Classification	Objective	Energy consumption	Hierarchy	Work center, segment, building	Energy form	Heat
	Function (product)	Storage, building structure, heating			Function (energy)	Consumptive usage

Initial situation	**Targeted situation**
Description	Principle and variants
Transport processes into or out of a building require doors and gates that lead to heat losses.	Transmission heat losses are reduced by appropriate choice of gates, *i.e.*, with a low heat transition coefficient. Information on leakage heat losses is provided by the gate manufacturer.
Cause	
The heat losses through gates are caused by three effects: First, the closed door causes transmission heat losses which are higher than of the adjacent walls. Secondly, ventilation heat losses occur through the closed gate due to leakages. Finally, heat losses are caused when gates are opened for a transport process. Whereas the first two categories are defined by the technical parameters of the gate, the latter one may be influenced during operation. Ventilation heat losses are influenced by temperature difference, air flow and cross-sectional area [1].	Ventilation heat losses may be reduced by technological or organizational changes (*e.g.*, reduce the opening times for transports). Consider the following measures [3-4]: – use air locks or air curtains, – use high-speed doors, – adjust the opening height to the size of the logistics vehicle, – seal sides and top and bottom edges of docking gates (*e.g.*, insulating panels), and – avoid simultaneous opening of opposite doors.
Relevance	Application area
Ventilation results in the highest heat losses as compared to the other categories, starting with an opening duration of three minutes per hour [2].	The measure is especially relevant for buildings with frequent and short-term transports (*e.g.*, high traffic of forklift trucks).

Benefit and effort	Expected benefit	A study on gate systems analyzed the saving potentials for a defined logistics building scenario [2]. The following table presents the identified saving effects in terms of heat energy demand, which refer to the comparison with a standard gate.			

Measure	Savings	Measure	Savings
High-speed doors	up to 30 %	Air locks	up to 88 %
Adjust gate opening to object size	up to 63 %	Unload truck in building	up to 90 %

Side effects
– Local or sudden temperature drops and draft effects may reduce the thermal comfort for employees.
– Since the energy loss depends on the inside temperature, the interdependency to EEM „Reduce average room temperature" needs to be considered.

Background	**External information sources**
Terms, definitions and theoretical explanations	Standardization
Transmission heat losses occur through closed surfaces (heat conduction) due to a temperature difference. *Ventilation heat losses* result from heat convection in fluid materials.	The standard DIN EN 12426 defines a classification for air permeability of doors and gates [5].

Sources
[1] Martin, H. (2011). Transport- und Lagerlogistik: Planung, Struktur, Steuerung und Kosten von Systemen der Intralogistik. 8th ed. Wiesbaden: Vieweg+Teubner.
[2] Hausladen, G.; Klimke, K.; Schneegans, J.; Rössel, T. (2013). Unterschiedliche Torsysteme in Industriegebäuden unter Berücksichtigung energetischer, bauklimatischer und wirtschaftlicher Aspekte.
[3] Energieverluste in Logistikzentren reduzieren – Richtige Planung sichert den Erfolg (2012). In: Fördern+heben 62 (3), pp. 40–42.
[4] Klimke, K. (2014). Wärmeverluste durch Staplerverkehr erfassen – Forschungsergebnisse zur Energiebilanz von Gebäudetoren in Lagerhallen. In: Fördern+heben 64 (5), pp. 80–83.
[5] Deutsches Institut für Normung (2000). DIN EN 12426 Industrial, commercial and garage doors and gates – Air permeability – Classification. Berlin: Beuth.

Number	H12	Date	XX.XX.XXXX

Figure 42: Energy efficiency measure implementation sheet for case study 2 – planning of logistics systems – Measure H12 "Reduce heat losses through in-bound and out-bound transport processes"

Industrial application example
Reduce heat losses through in-bound and out-bound transport processes

<table>
<tr><td rowspan="3">Object area</td><td>Industrial sector</td><td colspan="2">Motor vehicles and trailers</td><td>Location</td><td>Bruchsal, Baden-Württemberg, Germany</td></tr>
<tr><td>Company name</td><td colspan="2">John Deere</td><td>Number of employees</td><td>57,000 total; 700 at location [1]</td></tr>
<tr><td>Application area</td><td colspan="4">The measure was applied at a new spare parts center that expanded the existing site [2].</td></tr>
<tr><td rowspan="2">Implementation details</td><td>Measure description</td><td colspan="4">A new logistics center with an area of 16,000 m² with the dimensions 168 x 100 x 12 m was built [2]. Increasing energy efficiency is achieved through the following actions:
– Docks without thermal bridges
The gates are equipped with a 45 mm polyurethane hardfoam core to achieve a heat transmission coefficient of 0.51 W/m²K [3]. Furthermore, the gates reach up to the base of the driveway, which reduces typically occurding thermal bridges (*i.e.*, due to the loadable metal components in docks).
– Docking without drafts
Opening the gates automatically adjusts to the height of the vehicle. Air curtains at the inner side of the door leaves prevent cold air from entering the building. Furthermore, elastic covers made of 3 mm thick polyvinyl chloride are applied to the vehicle [2].
– Low standby consumption
A control unit for doors and docks reduces the standby energy consumption by switching off the power supply when it is not needed [2].

Overview on hall gates [4]   Gate with high thermal insulation [4]</td></tr>
<tr><td>Implementation year</td><td colspan="2">2011</td><td>Combination with other measures? ☐ No ☒ Yes:</td><td>Switch off equipment when not in use</td></tr>
<tr><td>Benefit and effort</td><td>Explanations</td><td colspan="4">The investment in the entire building was 23 million Euro [2]. Although values on the benefit are not available for this concrete case, the energy saving effect compared to conventional gates is generally around 10 % of the heat energy demand [5]. The electricity consumption of doors and gates is reduced by up to 70 % [5]. During standby, they consume as low as 0.5 W [2].</td></tr>
<tr><td rowspan="2">Sources</td><td colspan="5">[1]   John Deere (2016). Daten & Fakten 2016. URL: https://www.deere.de/de_DE/our_company/about_us/john_deere_deutschland/jd_deutschland.page
[2]   Mehr als grüner Lack (2012). In: Materialfluss, 43 (5), pp. 22–23.
[3]   Rein und raus mit Verstand (2013). In: Materialfluss, 44 (11), pp. 8–10.
[4]   Novoferm (n.d.). John Deere Bruchsal. URL: www.novoferm.com/de/industriegebaeude-_john_deere_bruchsal.html
[5]   So dicht wie möglich (2013). In: Materialfluss, 44 (4), pp. 26–28.</td></tr>
<tr><td>Number</td><td colspan="2">H12-E1</td><td>Date</td><td>XX.XX.XXXX</td></tr>
</table>

Figure 43: Application example for case study 2 – planning of logistics systems

The gates are equipped with a control to adjust the opening height to the size of the delivering logistics vehicle. Furthermore, air curtains are installed and an additional insulation is provided by a cover that is applied to the vehicle.

With the help of this supportive information, an individual assessment of each measure is conducted. For this assessment, only the measures that involve either the logistics planning or the AGV manufacturer are considered.[27] The assessment criteria are realizability and expected benefits.

Reducing the transported masses is possible through a proper selection of the vehicle (*e.g.*, lightweight) and through the material and design of the warehouse racks. A quantitative analysis would be necessary in order to evaluate the benefits of this measure.

Increasing the utilization rate of transport containers means to put as much material into the racks as possible considering the limitations of maximum load even though this increases the transport mass. In contrast to the explanations on the measure implementation sheet, the required warehouse area may be reduced in this case. Hence, the measure is expected to have a high benefit.

A general reduction of areas for transport is achieved by using AGVs instead of electric tractors. There are several influences that allow reducing the required space: First of all, the AGVs are smaller in size as compared to electric tractors. Furthermore, the special kind of AGV is capable of flexible movements, *i.e.*, turning around its own axis and moving in transverse and longitudinal directions. Moreover, the AGV lifts a warehouse rack as compared to a tractor that carries several racks as a trailer and, hence, needs more space for turning. Therefore, the width of the transport path is greatly reduced. Finally, the warehouse area is operated completely automated, which prevents the access of order picking staff, reducing the extra space for security purposes. Therefore, this measure is already considered by the goods-to-person order picking concept.

The ratio between load capacity and total mass is defined by the AGV manufacturer. Being aware of this measure sensitizes the logistics planning staff to pay attention to the fulfillment of this criterion when comparing and selecting a supplier.

A vehicle fleet management system helps to reduce energy consumption through assigning transport orders to the vehicles. In this case, a system is provided by the AGV manufacturer. In cooperation with the logistics planning and logistics management, enhancements on this system are possible.

The rolling resistance of logistics vehicles results from the combination of material and geometry of vehicle wheels and hall floor. The AGV manufacturer specifies requirements for the stability, wear resistance, flatness, and inclination of the floor. Additional effort is not put into the realization of this measure since it is expected to have a low benefit.

Reducing acceleration processes is especially important for heavy-weight and/or fast-driving vehicles. Since the vehicles operate slowly (maximum velocity 90 m/min) and are relatively light, the measure is prioritized low. However, logistics processes might be

---

[27] The industrial project has been conducted in cooperation with the logistics planning department at the early concept stage and with support of the AGV manufacturer.

optimized in terms of the length of the waiting queue at the pick station, since every start and stop increases energy consumption.

A reduction of the velocity may have a positive effect on energy consumption. However, the velocity of this vehicle is already relatively low. Furthermore, this would lead to a decreased productivity in terms of order-picks per hour, which is not intended with regard to the logistics performance.

The arrangement of the warehouse racks is an important task for logistics planning. This measure is realized by means of experience from the AGV manufacturer with other logistics centers that are equipped with this AGV system. It is important to consider the interrelations between the dimension of transport paths and layout planning (*e.g.*, allowing two-way transports). It should be noted that the transport distance does not influence the average power demand of a vehicle but it does reduce the cycle time for a transport order, which may reduce the number of required vehicles.

Splitting the conveyor fleet into light and heavy-weight vehicles is not relevant for the case study due to the quite homogeneous mass of the warehouse racks (200 to 600 kg).

The measures that address the batteries and their charging processes are neither relevant since they are applicable for lead acid batteries, whereas the vehicles of the case study are equipped with lead gel batteries. Nevertheless, the measures sensitize for a high efficiency of charging equipment. The charging strategy (*e.g.*, charge over night or charge at waiting times during the process) influences the efficiency as well as the battery service life.

The general measure to switch off equipment appropriately in this case refers to the vehicles, the battery charging system, the lighting, and the information and communication systems. The AGVs may be operated in a waiting mode and an idle mode (*e.g.*, at night). A possible measure is to switch them into idle mode even at wait times during the transport processes (*e.g.*, during picking at the station).

Heat losses for in-bound and out-bound transport processes are not considered so far since the focus was put on the logistics processes within the building. However, they give important indications for planning logistics buildings in general.

A reduction of the room temperature is relevant since only a small area of the entire logistics building contains manual workplaces. The logistics planning staff will check whether a decentralized heating system is possible.

Adjusting the room height to the requirements may be a promising measure to reduce heating energy consumption, which additionally supports a reduction of the height of lamps. The pick stations and the warehouse racks have a height of about 2 m. However, the building life cycle is much longer than the planning horizon for the use as a logistics building. In order to keep the flexibility for a different usage, a higher building might be preferred.

The logistics planning may influence the lighting energy consumption by splitting it into general and workspace lighting and by adjusting the illumination to the working task. It is planned to reduce the lighting in the automated areas to a minimum and equip the pick stations with a separate workspace lighting.

As a summary, the following measures are selected as most promising and are considered for the further development of the logistics concept:

- reduce areas for transport and providing material,

- arrange warehouse racks in order to minimize transport routes,

- increase utilization rate of transport containers,

- reduce number of acceleration processes,

- switch off equipment appropriately,

- split lighting into general and workspace lighting,

- reduce room temperature in automated areas,

- use decentralized heating systems in manual areas, and

- reduce room height.

5.5.5 Interpretation of Case Study Results

The case study demonstrates the method's functionality for an application in logistics planning. The relevant object systems contain the AGV fleet, the battery charging system, information and communication equipment, safety equipment, room heating, and lighting. Main parameters that determine the energy consumption of the entire system, are power, usage time, number of facilities (*e.g.*, vehicles), area, and volume. Further parameters specify the energy consumption of single systems (*e.g.*, driving distance influences the energy consumption of the AGV). The case study focuses on the logistics process in detail, while additionally considering the energy efficiency influential parameters of peripheral systems (*e.g.*, heating, lighting). This demonstrates that the degree of detail to apply the method may be adapted with respect to the individual purpose.

Seven actors are considered regarding their influence on energy efficiency, namely logistics planning, logistics management, order picking staff, maintenance staff, building planning, building services, and AGV manufacturer. Based on the analysis of the case study, 27 measures are generated, whereof 17 relate to the logistics process and 12 to the logistics building.[28] The measures are assigned to the actors logistics planning (14), logistics management (8), AGV manufacturer (7), building services (5), building planning (5), order picking staff (1), and maintenance (1). Some measures are assigned to several actors, which means that the interaction between actors is important for a measure's realization. For example, the implementation of a vehicle fleet management system may be realized by a cooperation between logistics planning and AGV manufacturer.

The assessment of measures is focused on the influence of logistics planning staff at an early concept stage. Hence, the results are narrowed down to 20 measures, which are assessed regarding their realizability and expected benefit with the help of EEMIS sheets. Based on the assessment, nine measures are identified that will be integrated into the logistics concept for goods-to-person order picking.

Each of the identified measures is assessed with a suitability degree. The calculated suitability degrees are not supposed to be compared between the two considered areas, *i.e.*, the logistics process and the logistics building.[29] Within the areas, the degrees reflect the relevance of measures. While the average suitability degree for all measures in the area of the logistics process is 37 %, the measures that are finally selected, achieve an average suitability degree of 49 %. This demonstrates that the calculated degree exhibits a tendency for the important measures in the use case. Nevertheless, an individual assessment is required as described in the procedure model.

As a result, the methodical approach supports the identification of measures and creates transparency on the influential parameters on energy consumption. Trade-offs, such as the influence of the utilization rate of warehouse racks on vehicle power demand (increasing) and required space (decreasing) are identified. The developed measure implementation sheets support the measure assessment.

5.6 Validation Results

Within the scope of the validation, the developed methodical approach is examined regarding its relevance and usefulness. The validation concept comprises a characteristic-based comparison of assessment criteria, the development of a prototype, and the method's application for two case studies.

The assessment of the evaluation criteria is used to validate the methodical approach against the research gap. This demonstrates that the method fulfills the requirements of identifying energy efficiency measures in a systematic optimization procedure while it is based on qualitative information. The object area and considerable energy flows are not limited to a specific set, as is the range of possible measures. Therefore, the method may

---

[28] Two of the measures may be assigned to either category.

[29] The suitability degree is a relative indicator that is scaled towards the maximum achievable calculated relevance (see Section 4.9). Hence, its value depends on the specified criteria of the planning task.

be applied to complex factory systems, covering a wide range of systems, processes, and energy carriers.

The implemented prototype contains a knowledge base of 200 energy efficiency measures and the EEMA algorithm. Graphical interfaces support the user in specifying the individual factory planning task, including the description of the technical object system, the project characteristics, and the actors' influences. When the calculations are performed, the result is a prioritized list of energy efficiency measures that are assigned to the respective actors. The prototype serves as a basis to conduct the case studies.

The case studies demonstrates the application of the method in two different planning areas. Whereas the first case study focuses on manufacturing processes, the second case study deals with a logistics system including a wider range of object systems. The main purpose of the first case study is to analyze the assignment of measures to the planning task and to the actors. The second case study creates transparency on parameters that influence the energy consumption of the entire system. The measure implementation support sheets of the second case study help to assess the energy efficiency measures, which enables the preparation of an implementation in more detail.

It can be concluded that the method is suitable to identify energy efficiency measures for planning tasks in various object areas. Both case studies show the data requirements to perform the method's application, *i.e.*, information on the planning project including the processes and systems as well as actors, their tasks, and their influences. For both case studies, the identified energy efficiency measures are evaluated with the suitability degree, *i.e.*, the degree, to which a measure fits the specified planning task.

The main advantage of the method lies in the focus on the optimization stage of an energy efficiency project, which is systematically supported. This means that the method is able to quickly generate solution approaches to increase the energy efficiency in a factory system during its planning phase. The required data is limited to qualitative information on the factory planning project. By this, parameters that may influence the energy efficiency are analyzed qualitatively, which increases the transparency on the energy consumption structure. Furthermore, the influence of various actors on relevant parameters is described in a transparent way, which allows to capture the interdisciplinary project work.

A critical reflection of the validation results shows that the identified energy efficiency measures are on an intermediate level of detail. This means that there is still the need to transfer the identified EEM to the concrete planning task. However, the gap between the methodical results and a practical application are lower as, for example, when using general compilations of energy efficiency principles as described in Section 3.5.3 (*e.g.*, reduce process losses). Furthermore, the method does not directly lead to a monetary assessment since the information provided by the measure implementation support is mainly qualitative and the contained quantitative information does not relate to the specific planning task.

The developed method follows a different general approach for energy efficiency projects (see Section 4.1). In the following, advantages and disadvantages of both approaches are discussed with reference to the second case study. A quantitative approach for this case contains modeling the logistics system, preparing and performing measurements, determining peripheral effects, and deducing improvement measures (Müller, Hopf & Krones, 2013; Hopf et al., 2016). The measurements contain acquiring the load profile of the automated guided vehicles and the power demand of the other equipment in various operational states (Hopf et al., 2016, pp. 34 f.). The results identify the high importance of reducing the required space, since the energy demand for heating and lighting has the highest influence (Hopf et al., 2016, p. 35). As a result, 18 energy efficiency measures are identified, whereof the saving effects may partially be quantified (Hopf et al., 2016, p. 36).

In contrast to that, the qualitative analysis as first step of the develop method only requires an understanding of the socio-technical system, that may be acquired during an initial meeting with the project partner. The effort for the analysis can be reduced through applying the qualitative method. The identification of measures is conducted almost automatically, but requires an existing comprehensive knowledge base of energy efficiency measures. Using the implemented prototype, 27 energy efficiency measures are generated. The qualitative assessment may reveal the necessity of further partial quantitative analyses. Gathering this reduced amount of data reduces the effort for quantitative measurements.

As a summary on the comparison between the developed qualitative method and a quantitative state of the art approach, the most important criterion is the trade-off between data acquisition effort and level of detail of the results. While quantitative approaches focus on the analysis and assessment, the qualitative method focuses on the optimization and provides support for the quick identification of solution approaches. The developed method tends to be less appropriate when an exact quantification of effects is required. Yet, the method provides valuable information on relevant measures in early planning or concept stages.

6 Conclusions

The last chapter summarizes the content and findings of the thesis and gives an outlook on potential further research work.

6.1 Summary

Energy efficiency is an essential goal for society, because the growing energy consumption leads to negative climatic effects. Since manufacturing industry has a high share at the energy consumption, increasing energy efficiency is an important objective with regard to economic and ecological aspects in the context of sustainability. Scientific studies show that there is a notable potential to increase energy efficiency. However, enterprises face barriers towards the implementation of these concepts that mainly arise from organizational (*e.g.*, lack of time) and information-related aspects (*e.g.*, lack of knowledge). Moreover, a relevant share of enterprises does not deduce improvement measures from energy efficiency analyses. Therefore, it is important to methodically support the identification of energy efficiency measures.

Within the strategy of sustainable manufacturing, the planning and operation of energy-efficient factories is particularly important. Factories are complex socio-technical systems that are characterized by a variety of technical facilities and personnel roles. The elements in a factory comprise the basic production factors equipment, material, and personnel and are interlinked by information, material, energy, capital, and personnel flows. Due to the system complexity, the identification of energy efficiency potentials is a complicated task, which requires methodical support. Factory planning is the systematic process to plan a factory and comprises a variety of tasks. Therefore, the tasks are usually conducted in interdisciplinary projects. Factory planning plays a crucial role for energy-efficient manufacturing since the energy consumption of a factory is mainly determined during early planning phases.

Increasing energy efficiency in factories is currently supported by various standards, methods, and instruments. The majority of the approaches follows the scheme of a quantitative analysis in order to create transparency on the energy consumption. However, the need to acquire energy consumption data hinders the application of methods in early planning phases. Additionally, these types of methods put an emphasis on the analysis rather than the direct improvement of a system. As a summary for the assessment, there is a lack of a methodical approach that provides a structured improvement of the entire factory system throughout the factory life cycle, *i.e.*, which may especially be applied in early planning phases.

Therefore, a method for identifying energy efficiency potentials has been developed in this thesis that aims at the assignment of energy efficiency measures (EEM) to a factory planning task. The method contains modeling concepts for both the planning task and the energy efficiency knowledge, a matching algorithm to perform the assignment, and a procedure model that describes the steps for the method's application.

The required input to apply the method is a qualitative model of the socio-technical factory system containing the domains object system, energy efficiency influential parameters, project characteristics, and actors. The model of the object system represents the technical resources within a factory and describes them in terms of their hierarchy and function. Energy efficiency influential parameters that affect a system's energy efficiency are analyzed qualitatively. The characteristics of the factory planning project and the energy efficiency project are acquired. Finally, the qualitative model contains a description of relevant actors, their tasks, and their influences on the corresponding parameters.

The energy efficiency measures are described by several classifying criteria in a knowledge base. The matching algorithm EEMA calculates a suitability degree that represents the fit between the planning task and the energy efficiency measures. The result of applying the algorithm is a list of the identified energy efficiency measures and their assignment to relevant actors. Furthermore, implementation support is provided in form of the EEMIS sheets. These sheets describe the classification, initial and targeted situation, benefit and effort of a measure, and may additionally provide industrial examples.

The validation shows that the method fulfills the initially defined requirements, especially in terms of data acquisition effort and applicability in factory planning projects. With regard to the practicability of the developed approach, a prototype has been implemented. It contains a knowledge base of 200 energy efficiency measures and supports the application of the methodical procedure with the help of a graphical user interface. The EEMA algorithm is part of this prototype, which means that a prioritized list of energy efficiency measures is generated based on the specifications of the planning task.

Afterwards, two case studies have been conducted to validate the usefulness of the method using the prototype. The first case focuses on the planning of a manufacturing process, whereas the second case addresses the planning of a logistics system. In both case studies, the planning situation has been analyzed qualitatively and energy efficiency measures have been identified. In the second case study, the measure implementation sheets have been analyzed additionally. By this, the energy efficiency measures have been qualitatively assessed in order to prepare a later implementation.

It is concluded that the method is suitable to identify energy efficiency potentials in planning projects. Moreover, the method's results create transparency on the influential parameters on energy efficiency including the relation to the respective actors. The effort to acquire data is reduced by applying the method as compared to quantitative state of the art approaches. The two case studies are characterized by different conditions; yet, the degree of detail of the method may be adjusted according to the requirements.

In summary, the developed method provides an approach to identify energy efficiency measures for factory systems. It helps to create transparency on the possibilities to influence energy consumption of a factory and the roles of various actors to perform this task. Furthermore, information on measures is provided that supports project participants in generating energy-efficient planning solutions. The main advantage of the method lies in the focus on the optimization stage of an energy efficiency project, which is supported systematically by the quick identification of solution approaches. Only qualitative information on the factory planning project is required to apply the method. Furthermore, the influence of various actors on influential parameters is described in a transparent way, which allows to capture the interdisciplinary project work.

6.2 Outlook

Based on the results of this thesis, several aspects have been identified that may form the basis for future research and development work:

Support of Continuous Improvement Processes in Factory Management

The main purpose of the method is to support factory planning projects, although an application during factory management is possible as well. However, the procedure model is adjusted to a unique project task. A seamless integration into factory management would require integrating the individual modeling concepts into a different procedure that is compatible to continuous improvement processes. In this case, actors, their tasks, and influences need to be specified only once and the identified measures need to be assorted for stepwise implementation.

Quantitative Assessment of Energy Efficiency Measures

The method integrates a qualitative assessment of energy efficiency measures due to the fact that information is provided on the implementation sheets without direct reference to the analyzed system. The partial acquisition of quantitative data is a possible subsequent step as part of the implementation plan. By creating transparency on the energy efficiency influential parameters, the method supports the preparation of quantitative measurements.

Economic Assessment of Energy Efficiency Measures

The method provides information for a rough estimate of costs and benefits of a measure by means of the measure implementation support. Investment decisions in practice usually require a detailed cost-benefit-analysis. Hence, a possible extension of the method contains the interface to a monetary assessment. The challenge for this task is to transfer the qualitative information from the implementation sheets into quantitative information in order to support the assessment.

Integration of Indirect Supporting Measures

The method considers energy efficiency measures that have a direct effect on the parameters of the energy consumption. In general, measures may also contain indirect supporting measures, *i.e.*, approaches that improve the prerequisites for reducing energy consumption. For example, the implementation of an energy metering system does not directly influence the energy consumption. However, with the help of its information, appropriate energy reducing strategies may be deduced. The integration of indirect measures into the method would require a different approach for the assignment task than the attachment to the energy efficiency influential parameters.

Consideration of Environmental Objectives

The method aims at increasing the energy efficiency of an enterprise. In a wider context, sustainable manufacturing addresses further objectives, such as water consumption and greenhouse gas emissions. Measures towards these objectives could easily be integrated into the methodical approach. Consequently, this would entail an adaptation of the structure of the energy efficiency influential parameters.

Knowledge Base for Energy Efficiency Influential Parameters

The identification of energy efficiency influential parameters is an important component of the methodical approach. It is used to assign planning actors to the corresponding energy efficiency measures. Hence, the extension with new energy efficiency measures requires the use of clearly described models of energy efficiency influential parameters. A possible extension is to develop a knowledge base that provides influential parameters for various object systems.

Supply Chain Management

The method addresses the factory system as highest hierarchical level for energy efficiency improvements. However, logistics transports in a global supply chain lead to significant environmental impact. A generalization of the method to the level of a supply chain network is basically possible, since the general concepts of the socio-technical system are transferable. Minor methodical adjustments would be necessary and need to be tested with further case studies.

References

Abele, E.; Beckmann, B. (2012). Energieeffizienzsteigerung von Fabriken. In: ZWF Zeitschrift für wirtschaftlichen Fabrikbetrieb 107 (4), pp. 261–265.

Abele, E.; Eisele, C.; Schrems, S. (2012). Simulation of the Energy Consumption of Machine Tools for a Specific Production Task. In: Proceedings of the 19th CIRP International Conference on Life Cycle Engineering. Berkeley, California, pp. 233–237.

Aggteleky, B. (1987). Fabrikplanung – Werksentwicklung und Betriebsrationalisierung. Band 1: Grundlagen – Zielplanung – Vorarbeiten, Unternehmerische und systemtechnische Aspekte, Marketing und Fabrikplanung. 2nd ed. Munich, Vienna: Carl Hanser.

Aggteleky, B. (1990). Fabrikplanung – Werksentwicklung und Betriebsrationalisierung. Band 2: Betriebsanalyse und Feasibility-Studie, Technisch-wirtschaftliche Optimierung von Anlagen und Bauten. 2nd ed. Munich, Vienna: Carl Hanser.

Alter, R. (2013). Strategisches Controlling – Unterstützung des strategischen Managements. 2nd ed. Munich: Oldenbourg.

Altshuller, G. S. (1984). Creativity as an Exact Science – The Theory of the Solution of Inventive Problems. Boca Raton, Florida: CRC Press.

American Welding Society, AWS (1994). ANSI/AWS A3.0-94 Standard Welding Terms and Definitions.

Baets, W. R. J. (2005). Knowledge Management and Management Learning – Extending the Horizons of Knowledge-Based Management. New York, New York: Springer Science + Business Media.

Bahrami, A.; Valentine, D. T.; Aidun, D. K. (2015). Computational Analysis of the Effect of Welding Parameters on Energy Consumption in GTA Welding Process. In: International Journal of Mechanical Sciences 93, pp. 111–119.

Bahrs, J. (2007). Wissensmanagement in der Praxis – Ergebnisse einer empirischen Untersuchung. Berlin: GITO.

Balogun, V. A.; Mativenga, P. T. (2013). Modelling of Direct Energy Requirements in Mechanical Machining Processes. In: Journal of Cleaner Production 41, pp. 179–186.

Bamberg, G.; Coenenberg, A. G.; Krapp, M. (2012). Betriebswirtschaftliche Entscheidungslehre. 15th ed. Munich: Franz Vahlen.

Bandte, H. (2007). Komplexität in Organisationen – Organisationstheoretische Betrachtungen und agentenbasierte Simulation. Wiesbaden: Deutscher Universitäts-Verlag.

Bauernhansl, T.; Mandel, J.; Wahren, S.; Kasprowicz, R.; Miehe, R. (2014). Energieeffizienz in Deutschland – eine Metastudie: Analyse und Empfehlungen. Berlin, Heidelberg: Springer.

Bayerisches Landesamt für Umwelt (2009). Leitfaden für effiziente Energienutzung in Industrie und Gewerbe. URL: https://www.schwaben.ihk.de/blob/aihk24/produktmarken/innovation_und_umwelt/downloads/Leitfaden_Effiziente_Energien utzung_in_Industrie_und_Gewerbe-data.pdf (visited on Sept. 30, 2016).

Bayerisches Landesamt für Umweltschutz (2004). Druckluft im Handwerk. URL: http://www.lfu.bayern.de/energie/co2_minderung/doc/druckluft.pdf (visited on Sept. 30, 2016).

Becker, J.; Probandt, W.; Vering, O. (2012). Grundsätze ordnungsmäßiger Modellierung – Konzeption und Praxisbeispiel für ein effizientes Prozessmanagement. Berlin, Heidelberg: Springer.

Becker, J.; Rosemann, M.; Schütte, R. (1995). Grundsätze ordnungsmäßiger Modellierung. In: Wirtschaftsinformatik 37 (5), pp. 435–445.

Bellgran, M.; Säfsten, K. (2010). Production Development – Design and Operation of Production Systems. London: Springer.

Betrieblicher Umweltschutz in Baden-Württemberg, BUBW (2014). Eine Informationsplattform des Ministeriums für Umwelt, Klima und Energiewirtschaft. Lernende Energieeffizienznetzwerke – Initiative Energieeffizienz. URL: http://www.bubw.de/?lvl=6006 (visited on Sept. 30, 2016).

Bey, N.; Hauschild, M. Z.; McAloone, T. C. (2013). Drivers and Barriers for Implementation of Environmental Strategies in Manufacturing Companies. In: CIRP Annals – Manufacturing Technology 62 (1), pp. 43–46.

BKCASE Editorial Board (2014). The Guide to the Systems of Engineering Body of Knowledge (SEBoK), Version 1.3. URL: http://sebokwiki.org/w/downloads/SEBoKv1.3_full.pdf (visited on Sept. 30, 2016).

Blesl, M.; Kessler, A. (2013). Energieeffizienz in der Industrie. Berlin, Heidelberg: Springer.

Bogdanski, G.; Spiering, T.; Li, W.; Herrmann, C.; Kara, S. (2012). Energy Monitoring in Manufacturing Companies – Generating Energy Awareness through Feedback. In: 19th CIRP International Conference on Life Cycle Engineering. Berkeley, California, pp. 539–544.

Bogdanski, G.; Schönemann, M.; Thiede, S.; Andrew, S.; Herrmann, C. (2013). An Extended Energy Value Stream Approach Applied on the Electronics Industry. In: Emmanouilidis, C.; Taisch, M.; Kiritsis, D.: Advances in Production Management Systems. Competitive Manufacturing for Innovative Products and Services. Berlin, Heidelberg: Springer, pp. 65–72.

Böhner, J. (2013). Ein Beitrag zur Energieeffizienzsteigerung in der Stückgutproduktion. Doctoral thesis. University of Bayreuth.

Böhner, J.; Kübler, F.; Steinhilper, R. (2013). Assessment of Energy Saving Potentials in Manufacturing Operations. In: Proceedings of the 22nd International Conference on Production Research (ICPR). Iguassu Falls, Brazil.

Bortz, J.; Döring, N. (2006). Forschungsmethoden und Evaluation – für Human- und Sozialwissenschaftler. 4th ed. Heidelberg: Springer Medizin.

Böttger, U. (2010). Energieeffizienz: Das ernüchternde Ergebnis einer VDI-Umfrage – Spritzgießer müssen ihre Hausaufgaben machen. In: Industrie-Anzeiger 132 (4), p. 46.

Brecher, C.; Herfs, W.; Heyer, C.; Triebs, J. (2014). Maschinenelemente und Baugruppen. In: Neugebauer, R.: Handbuch Ressourcenorientierte Produktion. Munich, Vienna: Carl Hanser, pp. 495–525.

Browning, T. R. (2001). Applying the Design Structure Matrix to System Decomposition and Integration Problems: A Review and New Directions. In: IEEE Transactions on Engineering Management 48 (3), pp. 292–306.

Broy, M.; Kuhrmann, M. (2013). Projektorganisation und Management im Software Engineering. Berlin, Heidelberg: Springer.

Brüggemann, A. (2005). KfW-Befragung zu den Hemmnissen und Erfolgsfaktoren von Energieeffizienz in Unternehmen. URL: https://www.kfw.de/Download-Center/Kon zernthemen/Research/PDF-Dokumente-Sonderpublikationen/Sonderpublikation. pdf (visited on Sept. 30, 2016).

Bründl, A.; Deutsch, N.; Heidenreich, S.; Krüger, L.; Schneider, C.; Schulze, M. (2012). Erfolgsfaktoren eines „Ganzheitlichen Energiemanagements" (GEM). URL: http:// www.pwc.de/de_DE/de/energiewende/assets/energiemanagement-masnahmenmix- und-organisation-sind-entscheidend.pdf (visited on Sept. 30, 2016).

Bundesministerium für Wirtschaft und Technologie, BMWi (2014). Making more out of Energy – National Action Plan on Energy Efficiency. URL: http://www.bmwi.de/ English/Redaktion/Pdf/nape-national-action-plan-on-energy-efficiency,property= pdf,bereich=bmwi2012,sprache=en,rwb=true.pdf (visited on Sept. 30, 2016).

Bundesministerium für Wirtschaft und Technologie, BMWi (2015). Energiedaten – nationale und internationale Entwicklung. URL: http://bmwi.de/DE/Themen/Energie/ Energiedaten-und-analysen/Energiedaten/gesamtausgabe,did=476134.html (visited on Sept. 30, 2016).

Bundesverband der Energie- und Wasserwirtschaft, BDEW (2015). Energie-Info Industriestrompreise – Ausnahmeregelungen bei Energiepreisbestandteilen (Aktualisierte Fassung). URL: https://www.bdew.de/internet.nsf/id/23AB0D60851F2923C1257E 88002EFA3E/$file/BDEW_Energie-Info_Industriestrompreise_160715_final_ohne _AP.pdf (visited on Sept. 30, 2016).

Bunse, K.; Vodicka, M.; Schönsleben, P. (2011). Energiemanagement in der Produktion – Ansätze zum Überwachen und Steuern der Energieeffizienz. In: Industrie Management 27 (6), pp. 53–56.

Bürger, V. (2010). Quantifizierung und Systematisierung der technischen und verhaltensbedingten Stromeinsparpotenziale der deutschen Privathaushalte. In: Zeitschrift für Energiewirtschaft 34 (1), pp. 47–59.

Buschmann, M. (2013). Planung und Betrieb von Energiedatenerfassungssystemen. Wissenschaftliche Schriftenreihe des Institutes für Betriebswissenschaften und Fabriksysteme, Heft 97. Doctoral thesis. Technische Universität Chemnitz.

Cagno, E.; Worrell, E.; Trianni, A.; Pugliese, G. (2013). A Novel Approach for Barriers to Industrial Energy Efficiency. In: Renewable and Sustainable Energy Reviews 19, pp. 290–308.

Canas, A. J.; Carff, R.; Hill, G.; Carvalho, M.; Arguedas, M.; Eskridge, T. C.; Lott, J.; Carvajal, R. (2005). Concept Maps: Integrating Knowledge and Information Visualization. In: Tergan, S.-O.; Keller, T.: Knowledge and Information Visualization – Searching for Synergies. Berlin, Heidelberg: Springer, pp. 205–219.

Chan, E. K. H. (2014). Standards and Guidelines for Validation Practices: Development and Evaluation of Measurement Instruments. In: Zumbo, B. D.; Chan, E. K. H.: Validity and Validation in Social, Behavioral, and Health Sciences. Heidelberg, New York: Springer International Publishing, pp. 9–24.

Chen, D.; Heyer, S.; Seliger, G.; Kjellberg, T. (2012). Integrating Sustainability within the Factory Planning Process. In: CIRP Annals – Manufacturing Technology 61 (1), pp. 463–466.

Chryssolouris, G. (2006). Manufacturing Systems: Theory and Practice. 2nd ed. New York: Springer International Publishing.

Clarke, C. (2005). Automotive Production Systems and Standardisation – From Ford to the Case of Mercedes-Benz. Heidelberg: Physica.

Cleff, T. (2015). Deskriptive Statistik und Explorative Datenanalyse – Eine computer-gestützte Einführung mit Excel, SPSS und STATA. 3rd ed. Wiesbaden: Springer Fachmedien.

Council of the European Union (2007). Presidency Conclusions – Brussels, 8/9 March 2007. URL: http://www.consilium.europa.eu/ueDocs/cms_Data/docs/pressData/en/ec/93135.pdf (visited on Sept. 30, 2016).

Danilovic, M.; Browning, T. R. (2007). Managing Complex Product Development Projects with Design Structure Matrices and Domain Mapping Matrices. In: International Journal of Project Management 25 (3), pp. 300–314.

Danilovic, M.; Sandkull, B. (2005). The Use of Dependence Structure Matrix and Domain Mapping Matrix in Managing Uncertainty in Multiple Project Situations. In: International Journal of Project Management 23 (3), pp. 193–203.

Despeisse, M.; Ball, P.; Evans, S.; Levers, A. (2012). Industrial Ecology at Factory Level – a Conceptual Model. In: Journal of Cleaner Production 31, pp. 30–39.

Deutsche Bundesstiftung Umwelt, DBU (2008). Energie effizient – Klimaschutz in Industrie und Gewerbe. URL: https://www.dbu.de/phpTemplates/publikationen/pdf/0810081044374fe4.pdf (visited on Sept. 30, 2016).

Deutsche Energie-Agentur GmbH, dena (2010a). Ratgeber Druckluftsysteme für Industrie und Gewerbe. URL: http://www.dena.de/fileadmin/user_upload/Publikationen/Stromnutzung/Dokumente/Ratgeber_Druckluft_Industrie_und_Gewerbe.pdf (visited on Sept. 30, 2016).

Deutsche Energie-Agentur GmbH, dena (2010b). Ratgeber Elektrische Motoren in Industrie und Gewerbe – Energieeffizienz und Ökodesign-Richtlinie. URL: http://www.ihk-emden.de/blob/emdihk24/innovation/downloads/2351516/dena_Ratgeber_Elektische_Motoren_in_Industrie_und_Gewerbe-data.pdf (visited on Sept. 30, 2016).

Deutsche Energie-Agentur GmbH, dena (2010c). Ratgeber Fördertechnik für Industrie und Gewerbe. URL: https://industrie-energieeffizienz.de/energiekosten-senken/energieeffiziente-technologien/foerdertechnik/ (visited on Sept. 30, 2016).

Deutsche Energie-Agentur GmbH, dena (2012). Steigerung der Energieeffizienz mit Hilfe von Energieeffizienz-Verpflichtungssystemen. URL: http://www.dena.de/fileadmin/user_upload/Presse/studien_umfragen/Energieeffizienz-Verpflichtungssysteme/Studie_Energieeffizienz-Verpflichtungssysteme_EnEffVSys.pdf (visited on Sept. 30, 2016).

Deutsche Energie-Agentur GmbH, dena (2016). dena-Referenzprojekte: Herausragende Beispiele für effiziente Energienutzung. URL: http://industrie-energieeffizienz.de/

energiekosten-senken/referenzprojekte-best-practice/dena-referenzprojekte/ (visited on Sept. 30, 2016).

Deutsches Institut für Normung (2003a). DIN 4701 Energy Efficiency of Heating and Ventilation Systems in Buildings. Berlin: Beuth.

Deutsches Institut für Normung (2003b). DIN 8580 Manufacturing Processes – Terms and Definitions, Division. Berlin: Beuth.

Deutsches Institut für Normung (2005). DIN EN 14610 Welding and Allied Processes – Definitions of Metal Welding Processes. Berlin: Beuth.

Deutsches Institut für Normung (2008). DIN 276-1 Building Costs – Part 1: Building Construction. Berlin: Beuth.

Deutsches Institut für Normung (2009a). DIN EN 1011 Welding – Recommendations for Welding of Metallic Materials – Part 1: General Guidance for Arc Welding. Berlin: Beuth.

Deutsches Institut für Normung (2009b). DIN EN ISO 14001 Environmental Management Systems – Requirements with Guidance for Use. Berlin: Beuth.

Deutsches Institut für Normung (2010). DIN EN 15900 Energy Efficiency Services – Definitions and Requirements. Berlin: Beuth.

Deutsches Institut für Normung (2011a). DIN EN ISO 50001 Energy Management Systems – Requirements with Guidance for Use. Berlin: Beuth.

Deutsches Institut für Normung (2011b). DIN V 18599 Energy Efficiency of Buildings – Calculation of the Net, Final and Primary Energy Demand for Heating, Cooling, Ventilation, Domestic Hot Water and Lighting – Part 1: General Balancing Procedure, Terms and Definitions, Zoning and Evaluation of Energy Sources. Berlin: Beuth.

Deutsches Institut für Normung (2011d). DIN V 18599 Energy Efficiency of Buildings – Calculation of the Net, Final and Primary Energy Demand for Heating, Cooling, Ventilation, Domestic Hot Water and Lighting – Part 2: Net Energy Demand for Heating and Cooling of Building Zones. Berlin: Beuth.

Deutsches Institut für Normung (2011e). DIN V 18599 Energy Efficiency of Buildings – Calculation of the Net, Final and Primary Energy Demand for Heating, Cooling, Ventilation, Domestic Hot Water and Lighting – Part 4: Net and Final Energy Demand for Lighting. Berlin: Beuth.

Deutsches Institut für Normung (2011c). DIN V 18599 Energy Efficiency of Buildings – Calculation of the Net, Final and Primary Energy Demand for Heating, Cooling, Ventilation, Domestic Hot Water and Lighting – Part 8: Net and Final Energy Demand of Domestic Hot Water Systems. Berlin: Beuth.

Deutsches Institut für Normung (2012a). DIN EN 15978 Sustainability of Construction Works – Assessment of Environmental Performance of Buildings – Calculation Method. Berlin: Beuth.

Deutsches Institut für Normung (2012b). DIN EN 16212 Energy Efficiency and Savings Calculation – Top-Down- and Bottom-Up Methods. Berlin: Beuth.

Deutsches Institut für Normung (2012c). DIN EN 16231 Energy Efficiency Benchmarking Methodology. Berlin: Beuth.

Deutsches Institut für Normung (2012d). DIN EN 16247-1 Energy Audits – Part 1: General Requirements. Berlin: Beuth.

Deutsches Institut für Normung (2012e). DIN EN ISO 15614 Specification and Qualification of Welding Procedures for Metallic Materials – Welding Procedure Test – Part 1: Arc and Gas Welding of Steel and Arc Welding of Nickel and Nickel Alloys. Berlin: Beuth.

Deutsches Institut für Normung (2013). DIN 4108 Thermal Insulation and Energy Economy in Buildings – Part 2: Minimum Requirements to Thermal Insulation. Berlin: Beuth.

Deutsches Institut für Normung (2014). DIN EN ISO 5817 Welding – Fusion-welded Joints in Steel, Nickel, Titanium and their Alloys (Beam Welding Excluded) – Quality Levels for Imperfections. Berlin: Beuth.

Deutsches Institut für Normung (2015a). DIN EN 16646 Maintenance – Maintenance within Physical Asset Management. Berlin: Beuth.

Deutsches Institut für Normung (2015b). DIN EN ISO 9000 Quality Management Systems – Fundamentals and Vocabulary. Berlin: Beuth.

Deutsches Institut für Normung (2015c). DIN EN ISO 9001 Quality Management Systems – Requirements. Berlin: Beuth.

Deutsches Institut für Normung (2016). DIN EN ISO 13273 Energy Efficiency and Renewable Energy Sources – Common International Terminology – Part 1: Energy Efficiency. Berlin: Beuth.

Deutsches Institut für Urbanistik (2006). Was ist eigentlich Flächenkreislaufwirtschaft? URL: http://www.difu.de/publikationen/difu-berichte-42006/was-ist-eigentlich-flaechenkreislaufwirtschaft.html (visited on Sept. 30, 2016).

Die Bundesregierung (2002). Perspektiven für Deutschland – Unsere Strategie für eine nachhaltige Entwicklung. URL: https://www.bundesregierung.de/Content/DE/_Anlagen/Nachhaltigkeit-wiederhergestellt/perspektiven-fuer-deutschland-langfassung.pdf (visited on Sept. 30, 2016).

Dietmair, A.; Verl, A.; Eberspächer, P. (2011). Model-based Energy Consumption Optimisation in Manufacturing System and Machine Control. In: International Journal of Manufacturing Research 6 (4), pp. 380–401.

Dietmair, A.; Verl, A.; Wosnik, M. (2008). Zustandsbasierte Energieverbrauchsprofile – Eine Methode zur effizienten Erfassung des Energieverbrauchs von Produktionsmaschinen. In: wt Werkstatttechnik online 98 (7/8), pp. 640–645.

Dombrowski, U.; Kynast, J. F.; Aurich, R. (2012). Energieeffizienz in gewachsenen Strukturen – Ein Modell zur Nutzung von Energieeinsparpotenzialen in KMU. In: ZWF Zeitschrift für wirtschaftlichen Fabrikbetrieb 107 (9), pp. 595–598.

Dombrowski, U.; Riechel, C. (2013). Sustainable Factory Profile: A Concept to Support the Design of Future Sustainable Industries. In: Proceedings of the 11th Global Conference on Sustainable Manufacturing. Berlin, pp. 73–78.

Drucker, P. F. (1974). Neue Management-Praxis, Band 1: Aufgaben. Düsseldorf, Vienna: Econ.

Duflou, J. R.; Kellens, K.; Renaldi; Guo, Y.; Dewulf, W. (2012a). Critical Comparison of Methods to Determine the Energy Input for Discrete Manufacturing Processes. In: CIRP Annals – Manufacturing Technology 61 (1), pp. 63–66.

Duflou, J. R.; Sutherland, J. W.; Dornfeld, D. A.; Herrmann, C.; Jeswiet, J.; Kara, S.; Hauschild, M.; Kellens, K. (2012b). Towards Energy and Resource Efficient Manufacturing: A Processes and Systems Approach. In: CIRP Annals – Manufacturing Technology 61 (2), pp. 587–609.

Eberhard, D. B. (2009). Personalentwicklung als Erfolgsfaktor einer strategischen Neuausrichtung zum Anbieter komplementärer Produkte und Dienstleistungen. In: Zink, K. J.: Personal- und Organisationsentwicklung bei der Internationalisierung von industriellen Dienstleistungen. Heidelberg: Physica, pp. 79–100.

ecofabrik (2016). Identifizierung und Bewertung energetischer und ökologischer Potentiale in Industriewerken – Quickcheck-Ablauf. URL: http://www.ecofabrik.eu/cms/viewPage/view/id/7 (visited on Sept. 30, 2016).

Elkington, J. (1997). Cannibals with Forks: Triple Bottom Line of 21st Century Business. Oxford: Capstone.

EnergieAgentur Nordrhein-Westfalen (2010a). Beleuchtung – Potenziale zur Energieeinsparung. URL: https://broschueren.nordrheinwestfalendirekt.de/broschuerenservice/staatskanzlei / beleuchtung - potenziale - zur - energieeinsparung / 1282 (visited on Sept. 30, 2016).

EnergieAgentur Nordrhein-Westfalen (2010b). Drucklufttechnik – Potenziale zur Energieeinsparung. URL: https://broschueren.nordrheinwestfalendirekt.de/broschuere nservice/energieagentur/drucklufttechnik-potenziale-zur-energieeinsparung/1368 (visited on Sept. 30, 2016).

EnergieAgentur Nordrhein-Westfalen (2010c). Elektrische Antriebe – Potenziale zur Energieeinsparung. URL: https://broschueren.nordrheinwestfalendirekt.de/broschue renservice/energieagentur/elektrische-antriebe-potenziale-zur-energieeinsparung/ 1369 (visited on Sept. 30, 2016).

Engel, A. (2010). Verification, Validation, and Testing of Engineered Systems. Hoboken, New Jersey: John Wiley & Sons.

Engelmann, J. (2009). Methoden und Werkzeuge zur Planung und Gestaltung energieeffizienter Fabriken. Wissenschaftliche Schriftenreihe des Institutes für Betriebswissenschaften und Fabriksysteme, Heft 71. Doctoral thesis. Technische Universität Chemnitz.

Engelmann, J. (2013). Ansätze zur Energieeffizienzsteigerung einer Karosseriebauhalle. In: Fachtagung Energieeffiziente Fabrik in der Automobilproduktion. Munich.

Engelmann, J.; Strauch, J.; Müller, E. (2008). Energieeffizienz als Planungsprämisse – Ressourcen- und Kostenoptimierung durch eine energieeffizienzorientierte Fabrikplanung. In: Industrie Management 24 (3), pp. 61–63.

Eppler, M. J. (2006). A Comparison between Concept Maps, Mind Maps, Conceptual Diagrams, and Visual Metaphors as Complementary Tools for Knowledge Construction and Sharing. In: Information Visualization 5 (3), pp. 202–210.

Erlach, K. (2013). Energy Value Stream: Increasing Energy Efficiency in Production. In: Schuh, G.; Neugebauer, R.; Uhlmann, E.: Future Trends in Production Engineering. Berlin, Heidelberg: Springer, pp. 343–349.

Erlach, K.; Sheehan, E. (2014). Die CO_2-Wertstrom-Methode zur Steigerung von Energie- und Materialeffizienz in der Produktion. In: ZWF Zeitschrift für wirtschaftlichen Fabrikbetrieb 109 (9), pp. 655–658.

Erlach, K.; Westkämper, E. (2009). Energiewertstrom – Der Weg zur energieeffizienten Fabrik. Stuttgart: Fraunhofer.

Esswein, W. (1993). Das Rollenmodell der Organisation: Die Berücksichtigung aufbauorganisatorischer Regelungen in Unternehmensmodellen. In: Wirtschaftsinformatik 35 (6), pp. 551–561.

European Commission (2003). Commission Recommendation Concerning the Definition of Micro, Small and Medium-sized Enterprises. URL: http://eur-lex.europa.eu/legal-content/EN/TXT/HTML/?uri=URISERV:n26026 (visited on Sept. 30, 2016).

European Commission (2009a). Commission Regulation (EC) No 640/2009 of 22 July 2009 Implementing Directive 2005/32/EC of the European Parliament and of the Council with Regard to Ecodesign Requirements for Electric Motors. URL: http://eur-lex.europa.eu/legal-content/EN/TXT/PDF/?uri=CELEX:32009R0640 (visited on Sept. 30, 2016).

European Commission (2009b). Directive 2009/125/EC Framework for the Setting of Ecodesign Requirements for Energy-related Products. URL: http://eur-lex.europa.eu/legal-content/EN/TXT/PDF/?uri=CELEX:32009L0125 (visited on Sept. 30, 2016).

European Commission (2014a). Commission Regulation (EU) No 4/2014 of 6 January 2014 Amending Regulation (EC) No 640/2009 Implementing Directive 2005/32/EC of the European Parliament and of the Council with Regard to Ecodesign Requirements for Electric Motors. URL: http://eur-lex.europa.eu/legal-content/EN/TXT/PDF/?uri=CELEX:32014R0004 (visited on Sept. 30, 2016).

European Commission (2014b). Energy Prices and Costs in Europe. URL: https://ec.europa.eu/energy/sites/ener/files/publication/Energy%20Prices%20and%20costs%20in%20Europe%20_en.pdf (visited on Sept. 30, 2016).

European Commission (2016). 2030 Climate & Energy Framework. URL: http://ec.europa.eu/clima/policies/strategies/2030/index_en.htm (visited on Sept. 30, 2016).

European Commission – Institute for Prospective Technological Studies, IPTS (2016). Best Available Techniques Reference Documents. URL: http://eippcb.jrc.ec.europa.eu/reference/ (visited on Sept. 30, 2016).

European Parliament (2012). Directive 2012/27/EU of the European Parliament and of the Council of 25 October 2012 on Energy Efficiency, Amending Directives 2009/125/EC and 2010/30/EU and Repealing Directives 2004/8/EC and 2006/32/EC. URL: http://eur-lex.europa.eu/legal-content/EN/TXT/PDF/?uri=CELEX:32012L0027 (visited on Sept. 30, 2016).

European Union (2012). Energy Roadmap 2050. URL: http://ec.europa.eu/energy/sites/ener/files/documents/2012_energy_roadmap_2050_en_0.pdf (visited on Sept. 30, 2016).

Fahrenwaldt, H. J.; Schuler, V.; Twrdek, J. (2014). Praxiswissen Schweißtechnik – Werkstoffe – Prozesse – Fertigung. 5th ed. Wiesbaden: Springer Fachmedien.

Feldmann, K.; Schöppner, V.; Spur, G. (2014). Handbuch Fügen, Handhaben, Montieren. Munich: Carl Hanser.

Felix, H. (1998). Unternehmens- und Fabrikplanung – Planungsprozesse, Leistungen und Beziehungen. Munich: Carl Hanser.

Fischer, J.; Weinert, N.; Herrmann, C. (2015). Method for Selecting Improvement Measures for Discrete Production Environments Using an Extended Energy Value Stream Model. In: Procedia CIRP 26, pp. 133–138.

Fischer, S. (2013). Material Efficiency in Companies of the Manufacturing Industry: Classification of Measures. In: Proceedings of the 11th Global Conference on Sustainable Manufacturing. Berlin, pp. 103–108.

Fleiter, T.; Hirzel, S.; Worrell, E. (2012). The Characteristics of Energy-Efficiency Measures – a Neglected Dimension. In: Energy Policy 51, pp. 502–513.

Frahm, B.-J.; Gruber, E.; Mai, M.; Roser, A.; Fleiter, T.; Schlomann, B. (2010). Evaluation des Förderprogramms „Energieeffizienzberatung" als eine Komponente des Sonderfonds' Energieeffizienz in kleinen und mittleren Unternehmen (KMU) – Schlussbericht. URL: http://www.isi.fraunhofer.de/isi-wAssets/docs/e/de/publikationen/evaluation-foerderprogramm-energieeffizienzberatung.pdf (visited on Sept. 30, 2016).

Fresner, J.; Jantschgi, J.; Birkel, S.; Bärnthaler, J.; Krenn, C. (2012). The Theory of Inventive Problem Solving (TRIZ) as Option Generation Tool within Cleaner Production Projects. In: Journal of Cleaner Production 18 (2), pp. 128–136.

Gausemeier, J.; Plass, C. (2014). Zukunftsorientierte Unternehmensgestaltung – Strategien, Geschäftsprozesse und IT-Systeme für die Produktion von morgen. 2nd ed. Munich: Carl Hanser.

Gessler, M. (2014). Handbuch für die Projektarbeit, Qualifizierung und Zertifizierung – auf Basis der IPMA Competence Baseline Version 3.0. Nuremberg: GPM Deutsche Gesellschaft für Projektmanagement.

Götze, U. (2014). Investitionsrechnung – Modelle und Analysen zur Beurteilung von Investitionsvorhaben. 7th ed. Berlin, Heidelberg: Springer.

Götze, U.; Müller, E.; Meynerts, L.; Krones, M. (2013). Energy-Oriented Life Cycle Costing – An Approach for the Economic Evaluation of Energy Efficiency Measures in Factory Planning. In: Neugebauer, R.; Götze, U.; Drossel, W.-G.: Energetisch-wirtschaftliche Bilanzierung und Bewertung technischer Systeme – Erkenntnisse aus dem Spitzentechnologiecluster eniPROD. Chemnitz: Verlag Wissenschaftliche Scripten, pp. 249–272.

Grundig, C.-G. (2015). Fabrikplanung: Planungssystematik – Methoden – Anwendungen. 5th ed. Munich: Carl Hanser.

Günther, U. (2005). Methodik zur Struktur- und Layoutplanung wandlungsfähiger Produktionssysteme. Wissenschaftliche Schriftenreihe des Institutes für Betriebswissenschaften und Fabriksysteme, Heft 50. Doctoral thesis. Technische Universität Chemnitz.

Günthner, W. A.; Galka, S.; Tenerowicz, P. (2009). Roadmap für eine nachhaltige Intralogistik. In: Tagungsband 14. Wissenschaftliche Fachtagung "Sustainable Logistics". Magdeburg, pp. 205–219.

Haag, H. (2013). Eine Methodik zur modellbasierten Planung und Bewertung der Energieeffizienz in der Produktion. Stuttgart: Fraunhofer.

Haapala, K. R.; Zhao, F.; Camelio, J.; Sutherland, J. W.; Skerlos, S. J.; Dornfeld, D. A.; Jawahir, I. S.; Clarens, A. F.; Rickli, J. L. (2013). A Review of Engineering Research in Sustainable Manufacturing. In: Journal of Manufacturing Science and Engineering 135 (4).

Haberfellner, R.; de Weck, O.; Fricke, E.; Vössner, S. (2012). Systems Engineering – Grundlagen und Anwendung. 12th ed. Zurich: orell füssli.

Hälsig, A.; Kusch, M.; Mayr, P. (2012). New Findings On The Efficiency Of Gas Shielded Arc Welding. In: Welding in the World 56 (11), pp. 98–104.

Hälsig, A.; Mayr, P. (2013). Energetische Bilanzierung von Fügeprozessen. In: Neugebauer, R.; Götze, U.; Drossel, W.-G.: Energetisch-wirtschaftliche Bilanzierung und Bewertung technischer Systeme – Erkenntnisse aus dem Spitzentechnologiecluster eniPROD. Chemnitz: Wissenschaftliche Scripten, pp. 283–298.

Hälsig, A.; Mayr, P.; Kusch, M. (2016). Determination of Energy Flows for Welding Processes. In: Welding in the World 60 (2), pp. 259–266.

Haußer, F.; Luchko, Y. (2011). Mathematische Modellierung mit MATLAB® – Eine praxisorientierte Einführung. Heidelberg: Spektrum Akademischer Verlag.

Hedrick, T. E.; Bickman, L.; Rog, D. J. (1993). Applied Research Design – A Practical Guide. Newbury Park, California: Sage Publications.

Herrmann, C. (2009). Ganzheitliches Life Cycle Management: Nachhaltigkeit und Lebenszyklusorientierung in Unternehmen. Berlin, Heidelberg: Springer.

Hesselbach, J. (2012). Energie- und klimaeffiziente Produktion – Grundlagen, Leitlinien und Praxisbeispiele. Wiesbaden: Vieweg+Teubner.

Hessisches Ministerium für Wirtschaft, Verkehr und Landesentwicklung (2009). Praxisleitfaden Energieeffizienz in der Produktion. URL: http://upp-kassel.de/wp-content/uploads/2013/09/Praxisleitfaden-Energieeffizienz-in-der-Produktion.pdf (visited on Sept. 30, 2016).

Hirzel, S.; Sontag, B.; Rohde, C. (2011). Betriebliches Energiemanagement in der industriellen Produktion. URL: http://www.effizienzfabrik.de/de/themen-ressourceneffizienz/studie-betriebliches-energiemanagement-in-der-industriellen-produktion/906/ (visited on Sept. 30, 2016).

Hitchins, D. K. (2007). Systems Engineering – A 21st Century Systems Methodology. West Sussex, England: John Wiley & Sons.

Hitomi, K. (1996). Manufacturing Systems Engineering – a Unified Approach to Manufacturing Technology, Production Management, and Industrial Economics. 2nd ed. London: Taylor & Francis.

Hoffmann, K.-H.; Witterstein, G. (2014). Mathematische Modellierung – Grundprinzipien in Natur- und Ingenieurwissenschaften. Basel: Springer.

Holzmüller, H. H.; Bandow, G. (2010). Einleitung – Zur disziplinbedingten "Färbung" von Modellen in der Betriebswirtschaftslehre und den Ingenieurwissenschaften. In: Bandow, G.; Holzmüller, H. H.: "Das ist gar kein Modell!" – Unterschiedliche Modelle und Modellierungen in Betriebswirtschaftslehre und Ingenieurwissenschaften. Wiesbaden: Gabler, pp. VII–XIV.

Hopf, H. (2016). Methodik zur Fabriksystemmodellierung im Kontext von Energie- und Ressourceneffizienz. Wiesbaden: Springer Fachmedien.

Hopf, H.; Müller, E. (2013a). Modeling of Energy-Efficient Factories with Flow System Theory. In: Prabhu, V.; Taisch, M.; Kiritsis, D.: Advances in Production Management Systems. Sustainable Production and Service Supply Chains. Berlin, Heidelberg: Springer Verlag, pp. 135–142.

Hopf, H.; Müller, E. (2013b). Visualization of Energy – Energy Cards Create Transparency for Energy-Efficient Factories and Processes. In: Azevedo, A.: Advances in Sustainable and Competitive Manufacturing Systems. Springer International Publishing, pp. 1665–1676.

Hopf, H.; Krones, M.; Brigl, T.; Müller, E. (2016). Teilautomatisierte Kommissionierung mit Fahrerlosen Transportsystemen – Bewertung der Energieeffizienz intralogistischer Konzepte. In: 21. Magdeburger Logistiktage "Logistik neu denken und gestalten". Magdeburg, pp. 29–37.

Hülsmann, S.; Köpschall, M.; Neumann, R.; Ohmer, M.; Hobusch, G.; Ruppelt, E.; Doll, M.; Krichel, S.; Sawodny, O.; Elsland, R.; Hirzel, S.; Schröter, M.; Weißfloch, U.;

Blank, F.; Nguyen, Q. K.; Roth-Stetlow, J. (2012). EnEffAH – Energieeffizienz in der Produktion im Bereich Antriebs- und Handhabungstechnik. URL: http://www.eneffah.de/EnEffAH_Broschuere.pdf (visited on Sept. 30, 2016).

Hutchins, M. J.; Robinson, S. L.; Dornfeld, D. A. (2013). Understanding Life Cycle Social Impacts in Manufacturing: A Processed-based Approach. In: Journal of Manufacturing Systems 32 (4), pp. 536–542.

IG Metall Baden-Württemberg (2003). Tarifvertrag tariflicher Niveaubeispiele vom 16.09.2003 in der Metall- und Elektroindustrie. URL: http://www.bw.igm.de/tarife/tarifvertrag.html?id=2531 (visited on Sept. 30, 2016).

Institute of Electrical and Electronics Engineers, IEEE (1991). IEEE Standard Computer Dictionary – Compilation of IEEE Standard Computer Glossaries.

Intergovernmental Panel on Climate Change, IPCC (2015). Climate Change – Synthesis Report 2014. URL: https://www.ipcc.ch/pdf/assessment-report/ar5/syr/SYR_AR5_FINAL_full_wcover.pdf (visited on Sept. 30, 2016).

International Building Performance Simulation Association, IBPSA (2016). Building Energy Software Tools Directory. URL: http://apps1.eere.energy.gov/buildings/tools_directory/ (visited on Sept. 30, 2016).

International Energy Agency, IEA (2012). World Energy Outlook 2012. URL: http://www.iea.org/publications/freepublications/publication/WEO2012_free.pdf (visited on Sept. 30, 2016).

International Energy Agency, IEA (2014). CO_2 Emissions from Fuel Combustion – Highlights. URL: http://apps.unep.org/redirect.php?file=/publications/pmtdocuments/-CO2_Emissions_from_Fuel_Combustion_Highlights-2014CO2_Emissions_From_FuelCombustion_Highl.pdf (visited on Sept. 30, 2016).

Itasse, S. (2014). Aus Nachhaltigkeit können Unternehmen viel Profit schlagen. In: Maschinenmarkt 120 (10), pp. 20–22.

Jayal, A. D.; Badurdeen, F.; Dillon Jr., O. W.; Jawahir, I. S. (2010). Sustainable Manufacturing: Modeling and Optimization Challenges at the Product, Process and System Levels. In: CIRP Journal of Manufacturing Science and Technology 2 (3), pp. 144–152.

Junge, M. (2007). Simulationsgestützte Entwicklung und Optimierung einer energieeffizienten Produktionssteuerung. Kassel: kassel university press GmbH.

Kaierle, S.; Dahmen, M.; Güdükkurt, O. (2011). Eco-efficiency of Laser Welding Applications. In: Proceedings of the International Society for Optical Engineering (SPIE);

SPIE Eco-Photonics 2011: Sustainable Design, Manufacturing, and Engineering Workforce Education for a Green Future. Strasbourg, France.

Kamiske, G. F. (2015). Handbuch QM-Methoden – Die richtige Methode auswählen und erfolgreich umsetzen. 3rd ed. Munich: Carl Hanser.

Kampker, A.; Franzkoch, B.; Hilchner, R. (2011). Type-Oriented Factory Planning with Solution Space Management. In: 4th International Conference on Changeable, Agile, Reconfigurable and Virtual Production (CARV). Montreal, Canada, pp. 569–573.

Karcher, P.; Siemer, M. (2013). Betriebliches Energiemanagement in produzierenden Unternehmen Deutschlands 2013. Berlin: Fraunhofer-Institut für Produktionsanlagen und Konstruktionstechnik IPK.

Karlsson, C. (2009). Researching Operations Management. New York: Taylor & Francis.

Kasabov, N. K. (1996). Foundations of Neural Networks, Fuzzy Systems, and Knowledge Engineering. Cambridge, Massachussetts: The MIT Press.

Kastens, U.; Büning, H. K. (2014). Modellierung – Grundlagen und formale Methoden. 3rd ed. Munich: Carl Hanser.

Keßler, H.; Winkelhofer, G. (2004). Projektmanagement – Leitfaden zur Steuerung und Führung von Projekten. 4th ed. Berlin, Heidelberg: Springer.

Keller, T.; Tergan, S.-O. (2005). Visualizing Knowledge and Information: An Introduction. In: Tergan, S.-O.; Keller, T.: Knowledge and Information Visualization – Searching for Synergies. Berlin, Heidelberg: Springer, pp. 1–26.

Kettner, H.; Schmidt, J.; Greim, H.-R. (1984). Leitfaden der systematischen Fabrikplanung. Munich, Vienna: Carl Hanser.

Kettner-Reich, A. (2011). Energieeffiziente Blechbearbeitung durch optimierte Fertigungstechnik. In: Maschinenmarkt 117 (25), pp. 26–28.

Klug, F. (2010). Logistikmanagement in der Automobilindustrie – Grundlagen der Logistik im Automobilbau. Berlin, Heidelberg: Springer.

König, C.; Kreimeyer, M.; Braun, T. (2008). Multiple-Domain Matrix as a Framework for Systematic Process Analysis. In: 10th International Design Structure Matrix Conference. Stockholm, Sweden, pp. 231–244.

Kramer, S. (1990). Application of Concept Mapping to Systems Engineering. In: Proceedings of the IEEE International Conference on Systems, Man and Cybernetics. Los Angeles, California, pp. 652–654.

Krause, M.; Thiede, S.; Herrmann, C.; Butz, F. F. (2012). A Material and Energy Flow Oriented Method for Enhancing Energy and Resource Efficiency in Aluminium

Foundries. In: Proceedings of the 19th CIRP International Conference on Life Cycle Engineering. Berkeley, California, pp. 281–286.

Krcmar, H. (2015). Informationsmanagement. 6th ed. Berlin, Heidelberg: Springer.

Kreimeyer, M.; Lindemann, U. (2011). Complexity Metrics in Engineering Design – Managing the Structure of Design Processes. Berlin, Heidelberg: Springer.

Krönert, S.; Fischer, A.; Fischer, F.; Götze, U. (2013). Wirtschaftliche Analyse von Handlungsalternativen am Beispiel der energiesensitiven Koordination von Robotern in getakteten Fertigungsstraßen. In: Neugebauer, R.; Götze, U.; Drossel, W.-G.: Energetisch-wirtschaftliche Bilanzierung und Bewertung technischer Systeme – Erkenntnisse aus dem Spitzentechnologiecluster eniPROD. Chemnitz: Wissenschaftliche Scripten, pp. 397–415.

Krones, M.; Hopf, H.; Müller, E. (2014). Ermittlung von Energieeffizienzmaßnahmen für Planung und Betrieb von Logistiksystemen. In: Neugebauer, R.; Drossel, W.-G.: Innovations of Sustainable Production for Green Mobility – Proceedings of the 3rd International Chemnitz Manufacturing Colloquium 2014 – Proceedings Part 2. Chemnitz: Verlag Wissenschaftliche Scripten, pp. 499–509.

Krones, M.; Müller, E. (2014a). An Approach for Reducing Energy Consumption in Factories by Providing Suitable Energy Efficiency Measures. In: Procedia CIRP 17, pp. 505–510.

Krones, M.; Müller, E. (2014b). Structuring Energy Efficiency Measures in Manufacturing Industry. In: Proceedings of the 24th International Conference on Flexible Automation and Intelligent Manufacturing (FAIM). San Antonio, Texas, pp. 241–248.

Krones, M.; Müller, E. (2016). Identification of Energy Efficiency Measures during Factory Planning Processes. In: Proceedings of the 6th International Conference on Industrial Engineering and Operations Management (IEOM). Kuala Lumpur, Malaysia, pp. 1092–1103.

Krones, M.; Shrivastava, A.; Pfefferkorn, F. E.; Müller, E. (2014). Bewertung der Ressourceneffizienz von Fertigungsprozessen – Überblick über Methoden und praktische Anwendung. In: Tagungsband TBI'14 – 15. Tage des Betriebs- und Systemingenieurs. Chemnitz, pp. 227–236.

Krones, M.; Hopf, H.; Brigl, T.; Müller, E. (2016). Energieeffizienz in der Automobilindustrie – Bewertung eines Logistikkonzeptes für die teilautomatisierte Kommissionierung. In: ProductivITy 21 (1), pp. 19–22.

Krüger, W. (2007). Theoretische und empirische Beiträge zur Fabrikplanung unter dem Aspekt des demografischen Wandels. Wissenschaftliche Schriftenreihe des Institutes für Betriebswissenschaften und Fabriksysteme, Heft 62. Doctoral thesis. Technische Universität Chemnitz.

Kubota, F. I.; da Rosa, L. C. (2013). Identification and Conception of Cleaner Production Opportunities with the Theory of Inventive Problem Solving. In: Journal of Cleaner Production 47, pp. 199–210.

Kuhrke, B. (2011). Methode zur Energie- und Medienbedarfsbewertung spanender Werkzeugmaschinen. Doctoral thesis. Technische Universität Darmstadt.

Landherr, M.; Neumann, M.; Volkmann, J.; Jäger, J.; Kluth, A.; Lucke, D.; Rahman, O.-A.; Riexinger, G.; Constantinescu, C. (2013). Fabriklebenszyklusmanagement. In: Westkämper, E.; Spath, D.; Constantinescu, C.; Lentes, J.: Digitale Produktion. Berlin, Heidelberg: Springer, pp. 163–195.

Lehner, F. (2014). Wissensmanagement – Grundlagen, Methoden und Technische Unterstützung. 5th ed. Munich: Carl Hanser.

Lehnertz, A. (2009). Gesteigerte Schweißeffizienz. In: BLECH InForm 9 (4), pp. 28–29.

Li, W.; Kara, S. (2011). An Empirical Model for Predicting Energy Consumption of Manufacturing Processes: a Case of Turning Process. In: Journal of Engineering Manufacture 225 (9), pp. 1636–1646.

Liker, J. K. (2007). Der Toyota Weg – 14 Managementprinzipien des weltweit erfolgreichsten Automobilkonzerns. 3rd ed. Munich: FinanzBuch.

Linke, B. S.; Corman, G. J.; Dornfeld, D. A.; Tönissen, S. (2013). Sustainability Indicators for Discrete Manufacturing Processes Applied to Grinding Technology. In: Journal of Manufacturing Systems 32 (4), pp. 556–563.

Liu, L; Li, C; Shi, J (2014). Analysis of Energy Utilisation Efficiency in Laser-GTA Hybrid Welding Process. In: Science and Technology of Welding and Joining 19 (7), pp. 541–547.

Lohre, D.; Bernecker, T.; Gotthardt, R. (2011). Praxisleitfaden zur IHK-Studie "Grüne Logistik". URL: https://www.stuttgart.ihk24.de/blob/sihk24/presse/Publikationen/ Branchen/Praxisleitfaden_Gruene_Logistik-data.pdf (visited on Sept. 30, 2016).

Lü, X.; Lu, T.; Kibert, C. J.; Viljanen, M. (2015). Modeling and Forecasting Energy Consumption for Heterogeneous Buildings Using a Physical-statistical Approach. In: Applied Energy 144, pp. 261–275.

Martin, H. (2014). Transport- und Lagerlogistik: Planung, Struktur, Steuerung und Kosten von Systemen der Intralogistik. 9th ed. Wiesbaden: Springer Fachmedien.

Miller, M. E.; Colombi, J. M.; Tvaryanas, A. P. (2014). Human Systems Integration. In: Badiru, A. B.: Handbook of Industrial and Systems Engineering. Boca Raton, Florida: CRC Press, pp. 197–216.

Mishra, R. S.; Ma, Z.-Y. (2005). Friction Stir Welding and Processing. In: Materials Science and Engineering 50, pp. 1–78.

Mishra, R. S.; De, P. S.; Kumar, N. (2014). Friction Stir Welding and Processing – Science and Engineering. Cham, Switzerland: Springer International Publishing.

Mose, C.; Weinert, N. (2015). Process Chain Evaluation for an Overall Optimization of Energy Efficiency in Manufacturing – The Welding Case. In: Robotics and Computer-Integrated Manufacturing 34, pp. 44–51.

Müller, E.; Hopf, H.; Krones, M. (2013). Analyzing Energy Consumption for Factory and Logistics Planning Processes. In: Emmanouilidis, C.; Taisch, M.; Kiritsis, D.: Advances in Production Management Systems. Competitive Manufacturing for Innovative Products and Services. Berlin, Heidelberg: Springer, pp. 49–56.

Müller, E.; Krones, M.; Hopf, H. (2012). Analyse und Bewertung des Energieverbrauchs von Intralogistiksystemen. In: Neugebauer, R.; Götze, U.; Drossel, W.-G.: Energieorientierte Bilanzierung und Bewertung in der Produktionstechnik – Methoden und Anwendungsbeispiele, 2. Methodenworkshop der Querschnittsarbeitsgruppe Energetisch-wirtschaftliche Bilanzierung des Spitzentechnologieclusters eniPROD. Chemnitz, pp. 359–374.

Müller, E.; Krones, M.; Strauch, J. (2013). Methodical Approach to Identify Energy Efficiency Measures in Factory Planning Based on Qualitative Analysis. In: Azevedo, A.: Proceedings of the 23rd International Conference on Flexible Automation and Intelligent Manufacturing (FAIM). Porto, Portugal, pp. 1627–1637.

Müller, E.; Löffler, T. (2009). Improving Energy Efficiency in Manufacturing Plants – Case Studies and Guidelines. In: Proceedings of the 16th CIRP International Conference on Life Cycle Engineering. Cairo, Egypt, pp. 465–471.

Müller, E.; Engelmann, J.; Löffler, T.; Strauch, J. (2009). Energieeffiziente Fabriken planen und betreiben. Berlin, Heidelberg: Springer.

Müller, E.; Poller, R.; Hopf, H.; Krones, M. (2013a). Enabling Energy Management for Planning Energy-efficient Factories. In: Procedia CIRP 7, pp. 622–627.

Müller, E.; Krones, M.; Strauch, J.; Fischer, S.; Veit, M. (2013b). Energieeffizienz von Stetigförderern – Analyse und Bewertung von Maßnahmen in Planung und Betrieb. In: Tagungsband 18. Magdeburger Logistiktage "Sichere und Nachhaltige Logistik". Magdeburg, pp. 153–160.

Müller, E.; Putz, M.; Krones, M.; Franz, E.; Hopf, H.; Langer, T.; Poller, R.; Kollatsch, C.; Schumann, M.; Klimant, P.; Wittstock, V. (2014). Energy-sensitive Solutions in Factory Planning and Factory Operation. In: Neugebauer, R.; Drossel, W.-G.: Innovations of Sustainable Production for Green Mobility – Proceedings of the 3rd International Chemnitz Manufacturing Colloquium 2014 and the 3rd International Colloquium of the Cluster of Excellence eniPROD – Proceedings Part 2. Chemnitz: Verlag Wissenschaftliche Scripten, pp. 213–230.

Müller, F.; Cannata, A.; Stahl, B.; Taisch, M.; Thiede, S.; Herrmann, C. (2013c). Green Factory Planning – Framework and Modules for a Flexible Approach. In: Proceedings of APMS 2013 International Conference Advances in Production Management Systems: Sustainable Production and Service Supply Chains. Berlin, Heidelberg: Springer Verlag, pp. 191–198.

National Aeronautics and Space Administration, NASA (2007). Systems Engineering Handbook. URL: http://ntrs.nasa.gov/archive/nasa/casi.ntrs.nasa.gov/20080008301.pdf (visited on Sept. 30, 2016).

Nebl, T. (2011). Produktionswirtschaft. 7th ed. Munich: Oldenbourg.

Nedjah, N.; Macedo Mourelle, L. de (2005). Real-World Multi-Objective System Engineering. Hauppauge, New York: Nova Science Publishers.

Neugebauer, R.; Westkämper, E.; Klocke, F.; Kuhn, A.; Schenk, M.; Michaelis, A.; Spath, D.; Weidner, E. (2008). Untersuchung zur Energieeffizienz in der Produktion. URL: http://edok01.tib.uni-hannover.de/edoks/e01fb09/59008982X.pdf (visited on Sept. 30, 2016).

Neugebauer, R.; Rennau, A.; Schönherr, J.; Fischer, S.; Schellenberger, S. (2010). Energieeffizienzmaßnahmen: Die produzierende Industrie im Spannungsfeld zwischen Staat und Käufer. In: ZWF Zeitschrift für wirtschaftlichen Fabrikbetrieb 105 (9), pp. 796–801.

Niegemann, H. M.; Domagk, S.; Hessel, S.; Hein, A.; Hupfer, M.; Zobel, A. (2008). Kompendium multimediales Lernen. Berlin, Heidelberg: Springer.

North, K. (2016). Wissensorientierte Unternehmensführung – Wissensmanagement gestalten. 6th ed. Wiesbaden: Springer Fachmedien.

Novikov, A. M.; Novikov, D. A. (2013). Research Methodology – from Philosophy of Science to Research Design. Boca Raton, Florida: CRC Press.

Olimpo, G. (2011). Knowledge Flows and Graphic Knowledge Representations. In: Trentin, G.: Technology and Knowledge Flow: the Power of Networks. Cambridge, UK: Chandos Publishing, pp. 91–131.

Orloff, M. A. (2012). Modern TRIZ: A Practical Course with EASyTRIZ Technology. Berlin, Heidelberg: Springer.

Ortlieb, C. P.; von Dresky, C.; Gasser, I.; Günzel, S. (2013). Mathematische Modellierung – eine Einführung in zwölf Fallstudien. 2nd ed. Wiesbaden: Springer Fachmedien.

Ott, A.; Cramer, J. (2009). Optimierter Energieeinsatz verbessert die Ertragssituation. In: ZWF Zeitschrift für wirtschaftlichen Fabrikbetrieb 104 (11), pp. 1018–1023.

Partsch, H. (2010). Requirements-Engineering systematisch – Modellbildung für softwaregestützte Systeme. 2nd ed. Berlin, Heidelberg: Springer.

Pehnt, M.; Arens, M.; Wolfgang Eichhammer, M. D. und; Fleiter, T.; Gerspacher, A.; Idrissova, F.; Jessing, D.; Jochem, E.; Kutzner, F.; Lambrecht, U.; Lehr, U.; Lutz, C.; Paar, A.; Reitze, F.; Schlomann, B.; Seefeldt, F.; Thamling, N.; Toro, F.; Vogt, R.; Wenzel, B.; Wünsch, M. (2011). Energieeffizienz: Potenziale, volkswirtschaftliche Effekte und innovative Handlungs- und Förderfelder für die Nationale Klimaschutzinitiative – Endbericht des Projektes "Wissenschaftliche Begleitforschung zu übergreifenden technischen, ökologischen, ökonomischen und strategischen Aspekten des nationalen Teils der Klimaschutzinitiative". URL: https://www.ifeu.de/energie/pdf/NKI_Endbericht_2011.pdf (visited on Sept. 30, 2016).

Pfefferkorn, F. E.; Lei, S.; Jeon, Y.; Haddad, G. (2009). A Metric for Defining the Energy Efficiency of Thermally Assisted Machining. In: International Journal of Machine Tools & Manufacture 49 (5), pp. 357–365.

Phillips, D. H. (2016). Welding Engineering – An Introduction. Hoboken, New Jersey: John Wiley & Sons.

Pohl, C.; Schevalje, C.; Hesselbach, J. (2013). Energetische Simulation – Ein effektives Werkzeug zur Identifikation von Optimierungspotenzialen in Druckluftnetzen. In: ZWF Zeitschrift für wirtschaftlichen Fabrikbetrieb 108 (3), pp. 137–142.

Pohl, J. (2013). Adaption von Produktionsstrukturen unter Berücksichtigung von Lebenszyklen. Doctoral thesis. Technical University of Munich.

Probst, G.; Raub, S.; Romhardt, K. (2012). Wissen managen – wie Unternehmen ihre wertvollste Ressource optimal nutzen. 7th ed. Wiesbaden: Gabler.

Reinema, C.; Mersmann, T.; Nyhuis, P. (2011). ecofabrik^{TM} – Internetbasierte Analyse der Energieeffizienz – Ein Ansatz für die Bewertung und Gestaltung energieeffizienter Fabriken. In: Industrie Management 27 (6), pp. 9–12.

Reinhart, G.; Karl, F.; Krebs, P.; Reinhardt, S. (2010). Energiewertstrom – Eine Methode zur ganzheitlichen Erhöhung der Energieproduktivität. In: ZWF Zeitschrift für wirtschaftlichen Fabrikbetrieb 105 (10), pp. 870–875.

Reinhart, G.; Karl, F.; Krebs, P.; Maier, T.; Niehues, K.; Niehues, M.; Reinhardt, S. (2011). Energiewertstromdesign – Ein wichtiger Bestandteil zum Erhöhen der Energieproduktivität. In: wt Werkstatttechnik online 101 (4), pp. 253–260.

Riedel, F. (2013). Selection of Joining Technologies for the Car Body Manufacturing Depending on Energy and Resource Efficiency. URL: http://publica.fraunhofer.de/eprints/urn_nbn_de_0011-n-2446575.pdf (visited on Sept. 30, 2016).

Riege, C.; Saat, J.; Bucher, T. (2009). Systematisierung von Evaluationsmethoden in der Wirtschaftsinformatik. In: Becker, J.; Krcmar, H.; Niehaves, B.: Wissenschaftstheorie und gestaltungsorientierte Wirtschaftsinformatik. Heidelberg: Physica, pp. 69–86.

Roberts, S. J. F.; Ball, P. D. (2014). Developing a Library of Sustainable Manufacturing Practices. In: Procedia CIRP 15, pp. 159–164.

Rodríguez, M. T. T.; Andrade, L. C.; Bugallo, P. M. B.; Long, J. J. C. (2011). Combining LCT Tools for the Optimization of an Industrial Process: Material and Energy Flow Analysis and Best Available Techniques. In: Journal of Hazardous Materials 192 (3), pp. 1705–1719.

Ropohl, G. (2009). Allgemeine Technologie – eine Systemtheorie der Technik. 3rd ed. Karlsruhe: Universitätsverlag.

Salonitis, K.; Ball, P. (2013). Energy Efficient Manufacturing from Machine Tools to Manufacturing Systems. In: Procedia CIRP 7, pp. 634–639.

Sauer, A.; Losert, F.; Pawlowski, A.; Bernards, M. (2013). Energieeffizienz in der Fertigungstechnik – Ergebnisse einer Kurzstudie. In: PRODUCTIVITY Management 18 (4), pp. 35–38.

Schacht, M.; Mantwill, F. (2012). Unterstützung des Planungsprozesses im Karosseriebau durch Energieverbrauchssimulation. In: ZWF Zeitschrift für wirtschaftlichen Fabrikbetrieb 107 (4), pp. 207–211.

Scheer, A.-W. (2002). ARIS in der Praxis: Gestaltung, Implementierung und Optimierung von Geschäftsprozessen. Berlin, Heidelberg: Springer.

Schenk, M.; Wirth, S.; Müller, E. (2010). Factory Planning Manual – Situation-Driven Production Facility Planning. Berlin, Heidelberg: Springer.

Schenk, M.; Wirth, S.; Müller, E. (2014). Fabrikplanung und Fabrikbetrieb – Methoden für die wandlungsfähige, vernetzte und ressourceneffiziente Fabrik. 2nd ed. Berlin, Heidelberg: Springer.

Schepers, S. W.; Meyer, G.; Nyhuis, P.; Wulf, S. (2013). Ressourceneffizienz kontinuierlich entwickeln – Verbesserung der Ressourceneffizienz in bestehenden Produktionsbetrieben. In: wt Werkstatttechnik online 103 (4), pp. 264–268.

Schlesinger, M.; Lindenberger, D.; Lutz, C. (2014). Entwicklung der Energiemärkte – Energiereferenzprognose: Studie im Auftrag des Bundesministeriums für Wirtschaft und Technologie (Endbericht). URL: http://www.bmwi.de/BMWi/Redaktion/PDF/Publikationen/entwicklung-der-energiemaerkte-energiereferenzprognose-endbericht.pdf (visited on Sept. 30, 2016).

Schmid, C. (2004). Energieeffizienz in Unternehmen – Eine wissensbasierte Analyse von Einflussfaktoren und Instrumenten. Zurich: vdf Hochschulverlag AG.

Schmid, C.; Brakhage, A.; Radgen, P.; Layer, G.; Arndt, U.; Carter, J.; Duschl, A.; Lilleike, J.; Nebelung, O. (2003). Möglichkeiten, Potenziale, Hemmnisse und Instrumente zur Senkung des Energieverbrauchs branchenübergreifender Techniken in den Bereichen Industrie und Kleinverbrauch. URL: http://www.isi.fraunhofer.de/isi-wAssets/docs/x/de/publikationen/ISI_REN-Querschnitt.pdf (visited on Sept. 30, 2016).

Schmigalla, H. (1995). Fabrikplanung – Begriffe und Zusammenhänge. Munich: Carl Hanser.

Schönsleben, P. (2001). Integrales Informationsmanagement: Informationssysteme für Geschäfts-prozesse – Management, Modellierung, Lebenszyklus und Technologie. 2nd ed. Berlin, Heidelberg: Springer.

Schreiber, G.; Akkermans, H.; Anjewierden, A.; de Hoog, R.; Shadbolt, N. R.; van de Velde, W.; Wielinga, B. J. (1999). Knowledge Engineering and Management: The CommonKADS methodology. Cambridge, Massachussetts: The MIT Press.

Schröter, M.; Weißfloch, U.; Buschak, D. (2009). Energieeffizienz in der Produktion – Wunsch oder Wirklichkeit? Energieeinsparpotenziale und Verbreitungsgrad energie-effizienter Techniken. URL: http://publica.fraunhofer.de/eprints/urn:nbn:de:0011-n-1180778.pdf (visited on Sept. 30, 2016).

Schuh, G.; Stich, V. (2012). Produktionsplanung und -steuerung 2 – Evolution der PPS. 4th ed. Berlin, Heidelberg: Springer.

Schulte, H. (2009). Fabrikplanung organisieren und durchführen. In: Maschinenbau – Das Schweizer Industriemagazin 38, pp. 88–90.

Schulze, C. P.; Reinema, C.; Nyhuis, P. (2012). Planung der Struktur einer Fabrik als soziotechnisches System. In: wt Werkstatttechnik online 102 (4), pp. 211–216.

Schütze, J. (2009). Modellierung von Kommunikationsprozessen in KMU-Netzwerken – Grundlagen und Ansätze. Wiesbaden: Gabler.

Seefeldt, F.; Berewinkel, J.; Lubetzki, C. (2009). Energieeffizienz in der Industrie – Eine makroskopische Analyse der Effizienzentwicklung unter besonderer Berücksichtigung der Rolle des Maschinen- und Anlagenbaus (Endbericht). URL: http://www.

prognos.com/fileadmin/pdf/publikationsdatenbank/Prognos_Energieeffizienz_in_der_Industrie.pdf (visited on Sept. 30, 2016).

Seefeldt, F.; Wünsch, M.; Michelsen, C.; Baumgartner, W.; Ebert-Bolla, O.; Matthes, U.; Leypoldt, P.; Herz, T. (2006). Potenziale für Energieeinsparung und Energieeffizienz im Lichte aktueller Preisentwicklungen – Endbericht. URL: http://www.prognos. com/uploads/tx_atwpubdb/070831_Prognos_BMWI_Potenziale_fuer_Energieeinsp arung.pdf (visited on Sept. 30, 2016).

Seidlmeier, H. (2015). Prozessmodellierung mit ARIS® – Eine beispielorientierte Einführung für Studium und Praxis in ARIS 9. 4th ed. Wiesbaden: Springer Fachmedien.

Senft, S. (2012). Roboter energieeffizient steuern und programmieren. In: etz Elektrotechnik und Automation 133 (5), pp. 2–3.

Shrivastava, A.; Krones, M.; Pfefferkorn, F. E. (2015). Comparison of Energy Consumption and Environmental Impact of Friction Stir Welding and Gas Metal Arc Welding for Aluminum. In: CIRP Journal of Manufacturing Science and Technology 9, pp. 159–168.

Shrivastava, A.; Krones, M.; Pfefferkorn, F. E. (2017). Joining Processes. In: Sutherland, J.; Dornfeld, D. A.; Linke, B.: Energy Efficient Manufacturing with Applications. Accepted for publication. Beverly, Massachussetts: Scrivener.

Shrivastava, A.; Overcash, M.; Pfefferkorn, F. E. (2015). Prediction of Unit Process Life Cycle Inventory (UPLCI) Energy Consumption in a Friction Stir Weld. In: Journal of Manufacturing Processes 18, pp. 46–54.

Siemens AG (2016). SinaSave Energy Efficiency Tool – Energy Efficiency Tool for Energy-Efficient Drive Technology. URL: http://www.industry.siemens.com/drives/ global/en/engineering-commissioning-software/sinasave/Pages/Default.aspx (visited on Sept. 30, 2016).

Smith, L.; Ball, P. (2012). Steps Towards Sustainable Manufacturing through Modelling Material, Energy and Waste Flows. In: International Journal of Production Economics 140 (1), pp. 227–238.

Sproesser, G.; Chang, Y.-J.; Pittner, A.; Finkbeiner, M.; Rethmeier, M. (2015). Life Cycle Assessment of Welding Technologies for Thick Metal Plate Welds. In: Journal of Cleaner Production 108 (Part A), pp. 46–53.

Stachowiak, H. (1973). Allgemeine Modelltheorie. Vienna, New York: Springer.

Statistisches Bundesamt (2007). Gliederung der Klassifikation der Wirtschaftszweige. URL: https://www.destatis.de/DE/Methoden/Klassifikationen/GueterWirtschaftkl assifikationen/Content75/KlassifikationWZ08.html (visited on Sept. 30, 2016).

Statistisches Bundesamt (2014). Nachhaltige Entwicklung in Deutschland – Indikatorenbericht 2014. URL: https://www.destatis.de/DE/Publikationen/Thematisch/UmweltoekonomischeGesamtrechnungen/Umweltindikatoren/IndikatorenPDF_0230001.pdf?_blob=publicationFile (visited on Sept. 30, 2016).

Statistisches Bundesamt (2015). Produzierendes Gewerbe: Kostenstruktur der Unternehmen des Verarbeitenden Gewerbes sowie des Bergbaus und der Gewinnung von Steinen und Erden 2013. URL: https://www.destatis.de/DE/Publikationen/Thematis ch/IndustrieVerarbeitendesGewerbe/Strukturdaten/Kostenstruktur2040430137004. pdf?_blob=publicationFile (visited on Sept. 30, 2016).

Steinhilper, R.; Freiberger, S.; Kübler, F.; Böhner, J. (2012). Development of a Procedure for Energy Efficiency Evaluation and Improvement of Production Machinery. In: Proceedings of the 19th CIRP International Conference on Life Cycle Engineering. Berkeley, California, pp. 341–346.

Steinmüller, W. (1993). Informationstechnologie und Gesellschaft – Einführung in die Angewandte Informatik. Darmstadt: Wissenschaftliche Buchgesellschaft.

Steward, D. V. (1981). The Design Structure System: A Method for Managing the Design of Complex Systems. In: IEEE Transactions on Engineering Management 28 (3), pp. 71–74.

Stock, T. (2016). Methode der dualen Energiesignaturen zur Analyse von Produktionsprozessen und erweiterten Wertstrombetrachtung. Wissenschaftliche Schriftenreihe des Institutes für Betriebswissenschaften und Fabriksysteme, Heft 121. Doctoral thesis. Technische Universität Chemnitz.

Tang, D.; Zhu, R.; Tang, J.; Xu, R.; He, R. (2010). Product Design Knowledge Management Based on Design Structure Matrix. In: Advanced Engineering Informatics 24 (2), pp. 159–166.

Thalheim, B.; Nissen, I. (2015). Wissenschaft und Kunst der Modellierung – Kieler Zugang zur Definition, Nutzung und Zukunft. Berlin, Boston: De Gruyter.

Thamling, N.; Seefeld, F.; Glöckner, U. (2010). Rolle und Bedeutung von Energieeffizienz und Energiedienstleistungen in KMU – Endbericht. URL: http://www.prognos. com/fileadmin/pdf/publikationsdatenbank/Prognos_Rolle_und_Bedeutung_von_Energieeffizienz_und_Energiedienstleistungen_in_KMU.pdf (visited on Sept. 30, 2016).

The Lowell Center for Sustainable Production (2016). What Is Sustainable Production? URL: http://www.sustainableproduction.org/abou.what.php (visited on Sept. 30, 2016).

Thiede, S.; Bogdanski, G.; Herrmann, C. (2012). A Systematic Method for Increasing the Energy and Resource Efficiency in Manufacturing Companies. In: Procedia CIRP 2, pp. 28–33.

Thiede, S.; Posselt, G.; Herrmann, C. (2013). SME Appropriate Concept for Continuously Improving the Energy and Resource Efficiency in Manufacturing Companies. In: CIRP Journal of Manufacturing Science and Technology 6 (3), pp. 204–211.

Trianni, A.; Cagno, E.; Donatis, A. D. (2014). A Framework to Characterize Energy Efficiency Measures. In: Applied Energy 118, pp. 207–220.

Trommer, G. (2010). Schweißtechnik: Vor- und nachgelagerte Prozessschritte als Kostentreiber – Wie beim Schweißen wirklich gespart wird. In: Industrie-Anzeiger 132 (13), p. 62.

Ulrich, H. (1984). Management. Bern: Paul Haupt.

Ulrich, H.; Probst, G. J. B. (2001). Anleitung zum ganzheitlichen Denken und Handeln – ein Brevier für Führungskräfte. Bern: Paul Haupt.

United Nations Industrial Development Organization (2011). Industrial Development Report 2011 – Industrial Energy Efficiency for Sustainable Wealth Creation – Capturing Environmental, Economic and Social Dividends. URL: http://www.unido.org/fileadmin/user_media/Publications/IDR/2011/UNIDO_FULL_REPORT_EBOOK. pdf (visited on Sept. 30, 2016).

United States Census Bureau (2012). North American Industry Classification System. URL: http://www.census.gov/eos/www/naics/2012NAICS/2012_Definition_File.pdf (visited on Sept. 30, 2016).

United States Environmental Protection Agency (2016a). Overview of Greenhouse Gases. URL: http://www3.epa.gov/climatechange/ghgemissions/gases.html (visited on Sept. 30, 2016).

United States Environmental Protection Agency (2016b). Reduce, Reuse, Recycle. URL: http://www2.epa.gov/recycle (visited on Sept. 30, 2016).

van den Bosch, P. P. J.; van der Klauw, A. C. (1994). Modeling, Identification and Simulation of Dynamical Systems. Boca Raton, Florida: CRC Press.

VDI Zentrum Ressourceneffizienz GmbH (2011). Umsetzung von Ressourceneffizienz-Maßnahmen in KMU und ihre Treiber – Identifizierung wesentlicher Hemmnisse und Motivatoren im Entscheidungsprozess von KMU bei der Inanspruchnahme öffentlicher Förderprogramme zur Steigerung der Ressourceneffizienz. URL: http://www.vdi-zre.de/fileadmin/user_upload/downloads/studien/28-11-2011_Broschuere_Web.pdf (visited on Sept. 30, 2016).

Verein Deutscher Ingenieure (1998). VDI-Guideline 3922 Energy Consulting for Industry and Business. Berlin: Beuth.

Verein Deutscher Ingenieure (2005). VDI-Guideline 2884 Purchase, Operating and Maintenance of Production Equipment using Life Cycle Costing (LCC). Berlin: Beuth.

Verein Deutscher Ingenieure (2010). VDI-Guideline 2695 Calculation of Operating Cost for Diesel and Electrical Forklift Trucks. Berlin: Beuth.

Verein Deutscher Ingenieure (2011). VDI-Guideline 5200 Factory Planning – Part 1: Planning Procedures. Berlin: Beuth.

Verein Deutscher Ingenieure (2013). VDI-Guideline 3807 Characteristic Values for Energy and Water Consumption of Buildings – Part 1: Fundamentals. Berlin: Beuth.

Verein Deutscher Ingenieure (2014a). VDI-Guideline 3802 Air Conditioning Systems for Factories. Berlin: Beuth.

Verein Deutscher Ingenieure (2014b). VDI-Guideline 4075 Cleaner Production (PIUS) – Part 1: Basic Principles and Area of Application. Berlin: Beuth.

Verein Deutscher Ingenieure (2016). VDI-Guideline 4070 Sustainable Management in Small and Medium-Sized Enterprises – Part 1: Guidance Notes for Sustainable Management. Berlin: Beuth.

Vereinigung der Bayerischen Wirtschaft, vbw (2012). Energieeffizienz in der Industrie. URL: http://www.vbw-bayern.de/Redaktion-%28importiert-aus-CS%29/04_Downl oads/Downloads_2012/03_Wirtschaftspolitik/3.5-Energie/Studie-Energieeffizienz-in-der-Industrie.pdf (visited on Sept. 30, 2016).

Verl, A.; Westkämper, E.; Abele, E.; Dietmair, A.; Schlechtendahl, J.; Friedrich, J.; Haag, H.; Schrems, S. (2011). Architecture for Multilevel Monitoring and Control of Energy Consumption. In: Proceedings of the 18th CIRP International Conference on Life Cycle Engineering. Berlin, Heidelberg: Springer Verlag, pp. 347–352.

Vogel-Heuser, B. (2003). Systems Software Engineering – Angewandte Methoden des Systementwurfs für Ingenieure. Munich: Oldenbourg.

von Carlowitz, H. C. (2013). Sylvicultura oeconomica oder Haußwirthliche Nachricht und Naturmäßige Anweisung zur Wilden Baum-Zucht. Reprint of the 2nd ed. from 1732. Munich: oekom.

Watter, H. (2015). Hydraulik und Pneumatik – Grundlagen und Übungen – Anwendungen und Simulation. 4th ed. Wiesbaden: Springer Fachmedien.

Weber, W.; Kabst, R.; Baum, M. (2014). Einführung in die Betriebswirtschaftslehre. 9th ed. Wiesbaden: Springer Fachmedien.

Weinert, N. (2010). Vorgehensweise für Planung und Betrieb energieeffizienter Produktionssysteme. Stuttgart: Fraunhofer.

Wiendahl, H.-P. (1996). Grundlagen der Fabrikplanung. In: Eversheim, W.; Schuh, G.: Betriebshütte – Produktion und Management. 7th ed. Berlin, Heidelberg: Springer, pp. 9–1 –9–30.

Wiendahl, H.-P. (2014). Betriebsorganisation für Ingenieure. 8th ed. Munich: Carl Hanser.

Wiendahl, H.-P.; Reichardt, J.; Nyhuis, P. (2014). Handbuch Fabrikplanung – Konzept, Gestaltung und Umsetzung wandlungsfähiger Produktionsstätten. 2nd ed. Munich, Vienna: Carl Hanser.

Windisch, H. (2014). Thermodynamik – Ein Lehrbuch für Ingenieure. 5th ed. Munich: Oldenbourg.

Wirth, S.; Schenk, M.; Müller, E. (2011). Fabrikarten, Fabriktypen und ihre Entwicklungsetappen. In: ZWF Zeitschrift für wirtschaftlichen Fabrikbetrieb 106 (11), pp. 799–802.

Wolff, D.; Kulus, D.; Dreher, S. (2012). Simulating Energy Consumption in Automotive Industries. In: Bangsow, S.: Use Cases of Discrete Event Simulation – Appliance and Research. Berlin, Heidelberg: Shaker, pp. 59–86.

World Commission on Environment and Development (1987). Our Common Future. URL: http://www.un-documents.net/our-common-future.pdf (visited on Sept. 30, 2016).

Wysocki, R. K. (2011). Executive's Guide to Project Management – Organizational Processes and Practices for Supporting Complex Projects. Hoboken, New Jersey: John Wiley & Sons.

Zhao, H.; Magoulès, F. (2012). A Review on the Prediction of Building Energy Consumption. In: Renewable and Sustainable Energy Reviews 16 (6), pp. 3586–3592.

Zimmermann, H.-J. (2008). Operations Research – Methoden und Modelle. Für Wirtschaftsingenieure, Betriebswirte, Informatiker. 2nd ed. Wiesbaden: Vieweg.

Appendix

A1 Additional Material for Case Study 1

In Section 5.4.3, the influence of actors has been presented for the role "product designer". The influences of the other roles are specified in the following, *i.e.*, process engineer (Table A1), production engineer (Table A2), machine supervisor (Table A3), machine setter (Table A4), and machine operator (Table A5).

Table A1: Assignment of influences on parameters for the role "process engineer" in case study 1 – planning of welding processes

		Pre-processing	Welding equipment	Post-processing
Factory planning phases	Structure planning	E_{pre}	$E_{op}, P_{op}, t_{op}, v_{weld}, P_{id}, t_{id}, P_{sb}, t_{sb}$	E_{post}
	Detailed planning	-	-	-
Product life cycle phases	Product design	-	-	-
	Work preparation	E_{pre}	$E_{op}, P_{op}, t_{op}, v_{weld}, P_{id}, t_{id}, P_{sb}, t_{sb}$	E_{post}
	Production planning	-	-	-
	Product manufacturing	-	-	-

Table A2: Assignment of influences on parameters for the role "production engineer" in case study 1 – planning of welding processes

		Pre-processing	Welding equipment	Post-processing
Factory planning phases	Structure planning	-	-	-
	Detailed planning	E_{pre}	$E_{op}, P_{op}, t_{op}, P_{id}, t_{id}, P_{sb}, t_{sb}$	E_{post}
Equipment life cycle phases	Concept	E_{pre}	$E_{op}, P_{op}, t_{op}, P_{id}, t_{id}, P_{sb}, t_{sb}$	E_{post}
	Development	E_{pre}	$E_{op}, P_{op}, t_{op}, P_{id}, t_{id}, P_{sb}, t_{sb}$	E_{post}
	Manufacturing	-	-	-
	Installation	-	-	-
	Operation	-	-	-

Table A3: Assignment of influences on parameters for the role "machine supervisor" in case study 1 – planning of welding processes

		Pre-processing	Welding equipment	Post-processing
Equipment life cycle phases	Concept	-	-	-
	Development	-	-	-
	Manufacturing	-	-	-
	Installation	-	-	-
	Operation	E_{pre}	E_{op}, t_{op}, t_{id}	E_{post}

Table A4: Assignment of influences on parameters for the role "machine setter" in case study 1 – planning of welding processes

		Pre-processing	Welding equipment	Post-processing
Equipment life cycle phases	Concept	-	-	-
	Development	-	-	-
	Manufacturing	-	-	-
	Installation	-	-	-
	Operation	-	$E_{op}, P_{op}, v_{weld}, P_{id}$	-

Table A5: Assignment of influences on parameters for the role "machine operator" in case study 1 – planning of welding processes

		Pre-processing	Welding equipment	Post-processing
Equipment life cycle phases	Concept	-	-	-
	Development	-	-	-
	Manufacturing	-	-	-
	Installation	-	-	-
	Operation	-	E_{op}, t_{op}, t_{id}	E_{post}

A2 Additional Material for Case Study 2

Situation Analysis

In Section 5.5.2, the qualitative modeling has been focused on the automated guided vehicles (AGV) since they represent the main process in this case. In the following, the other subsystems are analyzed.

The AGV fleet summarizes the existing vehicles and fulfills the function of material transport. The main purpose of considering this hierarchical level separately is the influence of the number of vehicles n_{AGV}. Furthermore, the power levels of vehicles may be averaged depending on the proportion of time that the vehicles spend driving and charging. For example, the average power demand of the fleet may be different from the operating power of a single vehicle when some vehicles are charging. Figure A1 shows the functional model of the AGV fleet.

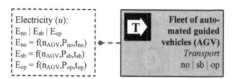

Figure A1: Functional model of the AGV fleet for case study 2 – planning of logistics systems

The battery charging system is used to charge the batteries of the logistics vehicles. Hence, its main function is to convert energy from a three-phase alternating current system into direct current. Besides, the unit causes energy losses in stand-by mode, *i.e.*, in times when no charging takes place. Therefore, the operational states stand-by and charging are distinguished. The energy consumption of a charging process can be calculated by (VDI 2695, p. 7):

$$E_{ch} = \frac{c_{bat} \cdot d_{dis} \cdot U_{cell} \cdot n_{cell} \cdot CF}{\eta_{ch}}, \tag{A1}$$

whereof c_{bat} is the nominal capacity of the battery, d_{dis} the degree of discharge before charging, U_{cell} the average voltage of a battery cell, n_{cell} the number of cells, CF the charging factor (depending on the battery system and charging method), and η_{ch} the efficiency of the charging system. Figure A2 shows the functional model of the battery charging system.

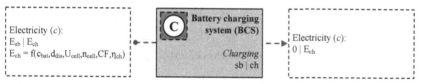

Figure A2: Functional model of the battery charging system for case study 2 – planning of logistics systems

The building system provides the environment for the logistics systems and consists of structural elements such as walls, floors, and windows. It is supplied with heat by the building services. For this use case, no further specific parameters need to be considered. Figure A3 depicts the functional model of the building system.

The room heating has the main function to generate heat by consuming heating gas. Although the heating system may have different usage states during the year, it is modeled as being operated continuously since the influence on the states is external, *i.e.*, depending on the weather conditions. In general, the power demand for heating depends on the volume of a room and the differences between inside and outside temperature. The heating energy demand results from the heat balance of a building that includes the effects of solar heat gains, internal heat gains (*e.g.*, through excess heat from machinery), transmission heat losses, and ventilation heat losses (DIN V 18599, Part 1, p. 28).

Figure A3: Functional model of the building system for case study 2 – planning of logistics systems

Solar heat gains depend on the area of transparent structural elements $A_{S,t}$ and the solar radiation I_S. Internal heat gains are usually accounted for by a specific coefficient per area h_P multiplied with the room area A, *e.g.*, 50 to 250 W/m² for a mechanical manufacturing shop (VDI 3802, p. 27). Transmission heat losses occur due to thermal conduction and convection on structural elements (*e.g.*, walls). They are influenced by the heat transition coefficient U, the area of the structural element A_S, the room temperature θ_i, and the outside temperature θ_o. Ventilation heat losses are influenced, besides the inside and outside temperature, by the room volume V and the air exchange rate n_{win} for continuous air exchange (DIN V 18599, Part 2, pp. 54 ff.), or by the opening cross-section A_{ae} and opening times t_{ae} for temporary air exchange, *e.g.*, through doors (Martin, 2014, pp. 324 f.). Figure A4 depicts the functional model of the room heating.

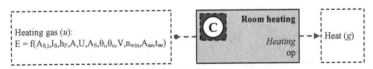

Figure A4: Functional model of the room heating for case study 2 – planning of logistics systems

The energy consumption for lighting depends on the power demand and the usage time. For describing the usage time, different zones within a building may be distinguished depending on the availability of daylight. For this case study, however, the influence of daylight is not addressed due to the generality of the logistics concept, *i.e.*, a concrete building scenario is not available. Thus, the usage time of the lighting system t_{op}^{ℓ} mainly depends on the production time (*e.g.*, shift times).

The lighting power is calculated according to (DIN V 18599, Part 4, pp. 26 f.):

$$P_\ell = A \cdot \frac{E_m}{MF \cdot \eta_L \cdot \eta_I \cdot \eta_R}, \tag{A1}$$

whereof A determines the area of the room, E_m the illumination intensity, MF a maintenance factor (*i.e.*, reduced illumination due to aging of the lamp), η_L the efficiency of the lamp, η_l the efficiency of the illuminant in the lamp, and η_R the lighting efficiency of the room. Figure A5 shows the functional model of the lighting.

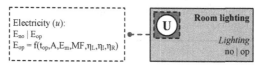

Figure A5: Functional model of the room lighting for case study 2 – planning of logistics systems

Both the information and communication equipment and the safety equipment, are utilizers of electricity in the states operating and non-operating. Each state is characterized by a defined power demand (P_{op}, P_{no}) and a usage time (t_{op}, t_{no}).

The actor model is supplemented by the description of the influences by logistics management, order picking staff, maintenance staff, building planner, building services staff, and AGV manufacturer.

The logistics management is responsible to control the vehicle fleet during operation. Hence, this role influences the energy consumption of the AGV in an operative way. The relevant parameters include the time spent in various operational states and the variables that describe the transport cycle, *i.e.*, transported mass, distance, number of acceleration processes, and share of empty trips. Table A6 summarizes the influence of the logistics management.

Table A6: Assignment of influences on parameters for the role "logistics management" in case study 2 – planning of logistics systems

Object system	Operation
AGV fleet	t_{no}, t_{sb}, t_{op}
AGV	$\alpha_{emp}, n_A, t_{no}, t_{sb}, t_{dr}, m_G, d_T$

The order picking staff works at the pick stations; their tasks are to pick material from the warehouse racks or to put material into the racks. Since this role does not affect the transport orders, the influence is limited to the usage times of peripheral equipment, *i.e.*, lighting and information and communication systems. In contrast to that, the heating system and safety equipment is assumed to be controlled centrally. Table A7 displays the influence matrix for the order picking staff.

The maintenance staff observes the technical condition of the equipment and, depending on that, conducts service and repair tasks. The influence on energy efficiency results from the fact that energy consumption of equipment may increase due to wear (Bayerisches Landesamt für Umwelt, 2009, p. 17). Thus, the maintenance staff may affect the parameters that are connected to the condition of equipment. Table A8 summarizes the influential opportunities for the maintenance staff.

Table A7: Assignment of influences on parameters for the role "order picking staff" in case study 2 – planning of logistics systems

Object system	Operation
Lighting	t_{op}
Information and communication equipment	t_{op}

Table A8: Assignment of influences on parameters for the role "maintenance staff" in case study 2 – planning of logistics systems

Object system	Maintenance
AGV	C_{rr}
Battery charging system	η_{ch}
Lighting	MF

The role building planning is responsible for the structural elements of the building. This includes the design, structure, and material of walls, windows, and floors. The planning tasks influence the heat balance of the building and, thereby, its heat energy demand. Furthermore, solar radiation gains influence the demand for artificial lighting. The concept planning includes defining the dimension of the building and specifying the construction material. In the detailed planning, details of the structural elements are determined, such as the orientation of windows and color of indoor walls. Table A9 summarizes the influential opportunities for the building planning staff.

Table A9: Assignment of influences on parameters for the role "building planning" in case study 2 – planning of logistics systems

Object system	Concept planning	Detailed planning
AGV	-	C_{rr}
Room heating	U, A, V, A_{ae}	$A_S, A_{S,t}, I_S$
Lighting	t_{op}	η_R

The building services staff plans and operates the building services, *i.e.*, heating, lighting, and air ventilation. Hence, this role may influence the aforementioned systems throughout their life cycle. The choice and technical design of the heating system affects the heat energy consumption. However, this is not explicitly included in the functional model of the heating since the model parameters are related to heat energy demand. Hence, the influence is generally described on the heat energy consumption E^{heat}. For the lighting system, the building services staff may influence the efficiency factors. Other influences, which may not be explicitly modeled, are possible. Table A10 summarizes the influential opportunities for the building services.

Table A10: Assignment of influences on parameters for the role "building services staff" in case study 2 – planning of logistics systems

Object system	Concept planning	Detailed planning	Operation
Room heating	E	E	E
Lighting	E, E_m	E, η_L, η_l	E

The AGV manufacturer closely works together with the logistics planning in order to specify the technical details of the vehicle. The main influence is on the technical characteristics of the vehicle. Additionally, the AGV system defines the required equipment for information/communication and safety. Finally, this actor provides the software for managing the logistics fleet, which is coordinated with the logistics management. Therefore, operating times and transport distances may be influenced during the operation of the system (*e.g.*, by assigning transport orders to vehicles). Table A11 summarizes the influential opportunities of the AGV manufacturer.

Table A11: Assignment of influences on parameters for the role "AGV manufacturer" in case study 2 – planning of logistics systems

Object system	Detailed planning
AGV	P_{no}, P_{sb}, m_{AGV}, a, C_{rr}
Battery charging system	c_{bat}, U_{cell}, n_{cell}, CF, η_{ch}
Information and communication equipment	P_{no}, P_{op}
Safety equipment	P_{no}, P_{op}

Energy Efficiency Measure Implementation Sheets

In Section 5.5.2, the energy efficiency measure implementation sheets have been presented exemplary for the measure H12 "Reduce heat losses through in-bound and out-bound transport processes". In the following, the measure implementation sheets for the other 26 measures are depicted.

Energy efficiency measure						
Reduce masses to be transported						
Objective	Energy consumption	Hierarchy	Work center		Energy form	Electricity, compressed air
Function (product)	Transport, storage	Function (energy)	Consumptive usage			

Initial situation	**Targeted situation**
Description	Principle and variants
Besides the useful load of goods, additional masses are moved during transport processes. These result in additional energy consumption.	Reducing the mass may be achieved by: – choosing different technologies (*e.g.*, electric tractors instead of forklift trucks), – using lightweight components (*e.g.*, roll conveyors made of plastics instead of metal), and – reducing packaging.
Cause	
Additional masses may be caused by: – containers and packaging material and – movable components of logistics systems (*e.g.*, counterweight of forklift trucks).	
Approaches to analyze initial situation	Application area
A helpful indicator is the ratio between own load and useful load capacity. This ratio should be minimized.	The mass needs to be reduced especially for dynamic movements [2]. Hence, the measure is important for uncontinuous conveyors, which are characterized by frequent changes in the direction of the movement [1].
Benchmark	
The share of energy consumption to move a conveyor without load is approx. 80 % for a forklift truck and more than 90 % for a storage and retrieval machine [1].	

The theoretical benefit can be calculated by the effect on the kinetic energy:

Expected benefit

$$E = \frac{1}{2}mv^2$$

Therefore, a reduction of the mass directly leads to the same proportion of energy savings.

[1] Furmans, K.; Linsel, P. (2011). Leichtbau bei Unstetigförderern durch Einsatz moderner Werkstoffe. In: Logistics Journal.
[2] Hülsmann, S.; Köpschall, M.; Neumann, R.; Ohmer, M.; Hobusch, G.; Ruppelt, E.; Doll, M.; Krichel, S.; Sawodny, O.; Elsland, R.; Hirzel, S.; Schröter, M.; Weißfloch, U.; Blank, F.; Nguyen, Q. K.; Roth-Stetlow, J. (2012). EnEffAH – Energieeffizienz in der Produktion im Bereich Antriebs- und Handhabungstechnik. URL: http://www.eneffah.de/EnEffAH_Broschuere.pdf (visited on September 30, 2016).

Number	T2	Date	XX.XX.XXXX

Figure A6: Energy efficiency measure implementation sheet for case study 2 – planning of logistics systems – Measure T2 "Reduce masses to be transported"

Energy efficiency measure					
Increase ratio between load capacity and total mass					

	Objective	Energy consumption	Hierarchy	Work center	Energy form	Electricity, compressed air
Classification	Function (product)	Transport, storage	Function (energy)	Consumptive usage		

Initial situation	**Targeted situation**
Description	Principle and variants
The energy that is required to transport an object depends on its mass. Hence, the energy efficiency of a logistics process is influenced by the ratio between the own mass and the load capacity.	An example for a favourable ratio is the multi shuttle technology in warehouses: One shuttle weighs 60 kg and may carry goods up to 40 kg. On the other hand, a storage and retrieval machine for small load carriers weighs 825 kg and has a capacity of 100 kg [2]. The lower energy consumption, which is caused by the reduced masses, allows to install smaller drives.
Benchmark	Another example is the comparison between a forklift truck and an electric tractor [3]: A case study in the automotive industry showed that the load capacity of an electric tractor is three times as high as the capacity of the forklift truck, whereas the own mass is 40 % lower.
The share of energy consumption to move a conveyor without load is approx. 80 % for a forklift truck and more than 90 % for a storage and retrieval machine [1].	
	Information need
	The own mass and load capacity is described in the data sheet of a logistics vehicle.

		The theoretical benefit can be calculated by the effect on the kinetic energy:
Benefit and effort	Expected benefit	$$E = \frac{1}{2}mv^2$$
		Therefore, a reduction of the mass directly leads to the same proportion of energy savings.

Background	**External information sources**
Terms, definitions and theoretical explanations	Standardization
The *rated load capacity* refers to defined standardized parameters (*e.g.*, lifting height of forklift trucks), since this influences the center of gravity of the good. Refer to load diagrams in order to determine the actual load capacity.	Data sheets of uncontinuous conveyors usually follow the structure as defined in VDI-guideline 2198 [4].

Sources	[1]	Furmans, K.; Linsel, P. (2011). Leichtbau bei Unstetigförderern durch Einsatz moderner Werkstoffe. In: Logistics Journal.
	[2]	Günthner, W. A.; Galka, S.; Tenerowicz, P. (2009). Roadmap für eine nachhaltige Intralogistik. In: Tagungsband zur 14. Wissenschaftlichen Fachtagung „Sustainable Logistics". Magdeburg, pp. 205-219.
	[3]	Veit, T.; Fischer, S.; Strauch, J.; Krause, A. (2010). Umsetzung logistischer Strategien unter Berücksichtigung der Energieeffizienz. In: Müller, E.; Spanner-Ulmer, B. (Hrsg.): Nachhaltigkeit in Planung und Produktion. Tagungsband zum 4. Symposium Wissenschaft und Praxis & 8. Fachtagung Vernetzt Planen und Produzieren.
	[4]	Verein Deutscher Ingenieure (2002). VDI-Guideline 2198 Type Sheets for Industrial Trucks. Berlin: Beuth.

Number	T6	Date	XX.XX.XXXX

Figure A7: Energy efficiency measure implementation sheet for case study 2 – planning of logistics systems – Measure T6 "Increase ratio between load capacity and total mass"

212

Appendix

Energy efficiency measure					
Use vehicle fleet management					
Objective	Energy consumption	Hierarchy	Work center, segment	Energy form	Electricity, diesel
Function (product)	Transport	Function (energy)	Consumptive usage		

Initial situation	Targeted situation
Description	Principle and variants
When several floor conveyors are used, a coordination of their operation is required. This includes determining the size of the fleet and controlling the operational states of each vehicle (*e.g.*, charging, operating).	A vehicle fleet management system acquires usage data of floor conveyors. This may include operating time, distance travelled, capacity of the battery, battery charging cycles, acceleration and braking processes, motor speed, frequency of collisions, or other parameters. With this data, the system may support the assignment between transport orders and vehicles. The goal is to reduce transport distances, which saves energy and may also reduce the number of required vehicles [3]. In general, a vehicle fleet management may help to:
Relevance	− optimize usage of conveyors and increase their service lifetime, − optimize usage of batteries and increase their service lifetime, and − reduce downtimes.
Reducing operational costs of floor conveyors is important since the investment causes 10 to 15 % of the life cycle costs [1]. Furthermore, the using phase accounts for 80 % of the greenhouse gases emitted during the life cycle of a forklift truck [2].	Implementation
	A variety of fleet management systems is available from manufacturers of floor conveyors, which may be used as a customer-specific service.

Expected benefit	The energy savings achieved through a vehicle fleet management are estimated as 20 % [4].	
Side effects	A vehicle fleet management increases productivity and safety, and reduces the operational costs of floor conveyors.	

[1] Dreier, J.; Hoppe, A.; Wehking, K.-H. (2012). Lebenszykluskosten ermitteln – Forschungsbedarf. In: Hebezeuge Fördermittel, 52 (9), pp. 476-478.
[2] *** (2012). Die CO_2-Spur des Staplers. In: LOG.Kompass, 8 (9).
[3] Günthner, W. A.; Galka, S.; Tenerowicz, P. (2009). Roadmap für eine nachhaltige Intralogistik. In: Tagungsband zur 14. Wissenschaftlichen Fachtagung „Sustainable Logistics". Magdeburg, pp. 205-219.
[4] Bayerisches Landesamt für Umwelt (2014). Logistik – Fuhrparkmanagement und Flottenmanagement. URL: http://www.izu.bayern.de/praxis/detail_praxis.php?pid=0203010100341 (visited on September 30, 2016).

Number	T8	Date	XX.XX.XXXX

Figure A8: Energy efficiency measure implementation sheet for case study 2 – planning of logistics systems – Measure T8 "Use vehicle fleet management"

Energy efficiency measure					
Increase utilization rate of transport containers					
Objective	Energy consumption	Hierarchy	Work center, segment	Energy form	Electricity, diesel
Function (product)	Transport	Function (energy)	Consumptive usage		

Initial situation	Targeted situation
Description	Principle and variants
A low utilization rate results in a low energy efficiency since the energy consumption for moving the empty conveyor is assigned to the utilization rate [1]. Furthermore, a low utilization rate increases the number of required transports as well as the required storage area for the containers.	The load capacity of transport containers needs to be adjusted to the requirements of a logistics process. If this is the case, transport orders should be summarized in order to increase the utilization rate of the containers.
	Application area
Benchmark	The measure is especially important for vehicles with a relatively high energy consumption during empty trips. In this case, the additional mass of the material does not have a high effect on the energy consumption.
The utilization rate and the proportion of empty trips depend on the type of the good: Whereas bulk products (*i.e.*, mass is the limiting factor to determine the transport) have a high utilization rate, light goods (*i.e.*, volume is the limiting factor) have a utilization rate of 30 to 40 % [2].	Information need
	Information that is helpful to assess the benefits of the measure includes the capacity of containers, the average mass/volume of goods, and the variations in the utilization rate. Furthermore, the energy demand for transports with varying masses should be known.

	Side effects	A higher utilization rate increases the transport mass and may conflict with the EEM „Reduce masses to be transported". The trade-off needs to be analyzed individually. Furthermore, this measure may lead to a reduced space demand (see EEM „Reduce areas for transport and providing material").

Background	External information sources
Terms, definitions and theoretical explanations	Standardization
The *utilization rate* may be expressed as ratio between actual load and rated load in terms of mass or volume. The *energy efficiency* of a logistics process may be represented by the ratio of the energy consumption and the logistics performance (*e.g.*, product of mass and transport distance).	The standard EN 16258 describes the allocation of energy consumption to transport processes [3].

[1]	Dykhoff, H.; Souren, R. (2007). Nachhaltige Unternehmensführung – Grundzüge industriellen Umweltmanagements. Berlin, Heidelberg: Springer.
[2]	Schmied, M.; Knörr, W. (2012). Carbon Footprint – Teilgutachten „Monitoring für den CO₂-Ausstoß in der Logistikkette". URL: https://www.umweltbundesamt.de/sites/default/files/medien/461/publikationen/4306.pdf (visited on September 30, 2016)
[3]	Deutsches Institut für Normung (2013). DIN EN 16258 Methodology for Calculation and Declaration of Energy Consumption and GHG Emissions in Transport Services (Freight and Passengers). Berlin: Beuth.

Number	T9	Date	XX.XX.XXXX

Figure A9: Energy efficiency measure implementation sheet for case study 2 – planning of logistics systems – Measure T9 "Increase utilization rate of transport containers"

Energy efficiency measure					
Use Ri charging process for lead acid batteries					

Classification					
Objective	Energy consumption	Hierarchy	Work center	Energy form	Electricity
Function (product)	Transport, process technology			Function (energy)	Conversion

Initial situation	**Targeted situation**
Description	Principle and variants
The mainly applied charging process is the so-called IUI charging process [1]. This means that the charging current is selected depending on the capacity of the battery.	The Ri charging process does not apply a constant charging current; it rather controls the charging current depending on the internal resistance of the battery [1]. This is realized through measuring the internal resistance and determining a suitable charging voltage.
Cause	
The constant charging current leads to high charging losses, especially at those points of the characteristic line with a high internal resistance.	
Approaches to analyze initial situation	This results in the following charging process: In the beginning, the internal resistance is high, hence, the charging current is set low to reduce losses. Once the internal resistance decreases, the charging current is increased. By this method, the battery is provided with the necessary charging current.
The energy consumption of the charging device may be measured with a power logger. Measuring the energy that is charged into the battery requires a device to measure direct current (*e.g.*, battery controller).	
Benchmark	Application area
The charging factor of an IUI charging process usually lies between 1.12 and 1.25, whereas the Ri charging process achieves a charging factor between 1.05 and 1.12 [1].	The Ri charging process is applicable for lead acid batteries.

Benefit and effort		
Expected benefit	The efficiency of the charging process increases by 10 % as compared to traditional charging processes [1].	
Side effects	Charging losses cause battery warming, which may reduce the service life of the battery.	

Background	**External information sources**
Terms, definitions and theoretical explanations	Standardization
The *internal resistance* of a lead acid battery (R_i) is a battery-specific value which depends on age, temperature, and degree of discharging [1]. The *charging efficiency* (η_{ch}) is defined as the ratio between the electrical charge to the battery and the energy that is required from the charging device. Its reciprocal is the *charging coefficient* [2].	The standard EN 16796 describes how to determine the charging efficiency and the efficiency of a charging device [3].

Sources	
[1]	Fronius International GmbH (2015). Batterien effizient und schonend laden – Technische Hintergründe zum Ri-Ladeprozess von Fronius. In: Fördern+heben, 65 (7-8), pp. 22-24.
[2]	Fronius International GmbH (2013). Active Inverter Technology with Ri-charging process. URL: http://www.fronius.com/cps/rde/xbcr/SID-78E30885-520CA30C/fronius_international/BLS_Flyer_Ri_chargingprocess_EN_v01_Feb_2013_ab10_low_289825_snapshot.pdf (visited on September 30, 2016).
[3]	Deutsches Institut für Normung (2014). DIN EN 16796-1 Energy Efficiency of Industrial Trucks – Test Methods – Part 1: General (Draft). Berlin: Beuth.

Number	T11	Date	XX.XX.XXXX

Figure A10: Energy efficiency measure implementation sheet for case study 2 – planning of logistics systems – Measure T11 "Use Ri charging process for lead acid batteries"

Energy efficiency measure						
Use high frequency charging equipment with electrolyte circulation						
Classification	Objective	Energy consumption	Hierarchy	Work center	Energy form	Electricity
	Function (product)	Transport	Function (energy)	Conversion		

Initial situation	Targeted situation
Description	Principle and variants
Traditional chargers with 50 Hz technology do not follow a controlled charging characteristic.	High frequency chargers contain a microprocessor to control the charging process (*e.g.*, depending on the charge level of the battery).
Relevance	
A case study among eight logistics buildings demonstrated that one half uses 50 Hz transformer technology and the other half uses high frequency charging [1].	
Benchmark	Application area
Chargers with 50 Hz technology have a total efficiency of 56 % (device efficiency 80 % and charging efficiency 70 %). High frequency charging equipment has a total efficiency of 68 % (device efficiency 90 % and charging efficiency 75 %). [2]	The use of high frequency chargers exploits savings potentials when applied for small charging voltages (up to 48 V) or when it substitutes inefficient charging characteristics (*e.g.*, Wsa) [3].

Benefit and effort		
	Energy savings	15 % to 30 % [4-5]
	Expected benefit	A case study in the automotive industry revealed that the use of high frequency chargers may reduce charging losses from 35 % to 5 % [6]. When using forklift trucks, it is assumed to save 30 % of charging energy by using high frequency chargers with electrolyte circulation [5].
	Side effects	High frequency chargers have a lower reactive power. On the other hand, their service lifetime may be shorter; it depends on usage and maintenance and may last up to 15 years. [3]

Background	External information sources
Terms, definitions and theoretical explanations	Standardization
The *charging efficiency* (η_{ch}) is defined as the ratio between the electrical charge to the battery and the energy that is required from the charging device (electrochemical process). Its reciprocal is the *charging coefficient*. The *device efficiency* (η_D) results from converting the alternating current from the socket into direct current in the battery charger (electrical process). The product of the charging efficiency and the device efficiency is the *total efficiency* of the charging process. [2]	The standard EN 16796 describes how to determine the charging efficiency and the efficiency of a charging device [7]. The standard DIN 41772 defines basic charging characteristics [8]. The assignment between batteries and charging devices is supported by an information leaflet from the German Electrical and Electronic Manufacturers' Association [9].

Sources	
[1]	Günthner, W. A.; Hausladen, G.; Freis, J.; Vohlidka, P. (2014). Das CO₂-neutrale Logistikzentrum – Entwicklung von Handlungsempfehlungen für energieeffiziente Logistikzentren. URL: http://www.fml.mw.tum.de/fml/index.php?Set_ID=870&Download=Forschungsbericht_Das_CO2_neutrale_Logistikzentrum_IGF_398ZN (visited on September 30, 2016).
[2]	Fronius International GmbH (2013). Active Inverter Technology with Ri-charging process. URL: http://www.fronius.com/cps/rde/xbcr/SID-78E30885-520CA30C/fronius_international/BLS_Flyer_Ri_chargingprocess_EN_v01_Feb_2013_ab10_low_289825_snapshot.pdf (visited on September 30, 2016).
[3]	*** (2010). HF-Ladetechnik – Allein glückselig machend? In: Staplerworld, 8 (2), pp. 27-29.
[4]	Kohagen, J. (2010). Umweltbilanz für Logistikzentren. In: DVZ, 64 (51).
[5]	Kramer, J. (2012). Energieeinsparpotenziale rund um das Flurförderzeug. In: Hebezeuge Fördermittel 52 (4), pp. 186-188.
[6]	Veit, T.; Fischer, S.; Strauch, J.; Krause, A. (2010). Umsetzung logistischer Strategien unter Berücksichtigung der Energieeffizienz. In: Tagungsband zum 4. Symposium Wissenschaft und Praxis & 8. Fachtagung Vernetzt Planen und Produzieren, Chemnitz.
[7]	Deutsches Institut für Normung (2014). DIN EN 16796-1 Energy Efficiency of Industrial Trucks – Test Methods – Part 1: General (Draft). Berlin: Beuth.
[8]	Deutsches Institut für Normung (1979). DIN 41772 Semiconductor Rectifier Equipment – Shapes and Letter Symbols of Characteristic Curves. Berlin: Beuth.
[9]	Zentralverband Elektrotechnik- und Elektronikindustrie e. V., ZVEI (2004). Ladegerätezuordnung für Antriebsbatterien in geschlossener (PzS) und verschlossener (PzV) Ausführung. URL: http://www.zvei.org/Verband/Fachverbaende/Batterien/Documents/Merkblaetter/11%20Ladegeraetezuordnung%20fuer%20Antriebsbatterien%202004-04.pdf (visited on September 30, 2016).

Number	T13	Date	XX.XX.XXXX

Figure A11: Energy efficiency measure implementation sheet for case study 2 – planning of logistics systems – Measure T13 "Use high frequency charging equipment with electrolyte circulation"

Classification					
Energy efficiency measure **Reduce rolling resistance of logistics equipment**					
Objective	Energy consumption	Hierarchy	Work center, component	Energy form	Electricity
Function (product)	Transport	Function (energy)	Consumptive usage		

Initial situation	**Targeted situation**
Description	Principle and variants
The energy consumption for a transport task with a floor conveyor results from the force to overcome the acceleration resistance, rolling resistance, and gradient resistance. When considering a continuous conveyor, the energy consumption is influenced by lifting resistance and friction resistance. Hence, reducing the rolling resistance leads to a lower energy consumption. [1]	Influencing the rolling resistance may be achieved by [4]: – coating surfaces, – selecting proper material pairing for roll conveyors, – using low-friction wheels for floor conveyors, and – using flat and leveled surfaces.
Cause	Application area
The rolling resistance is a complex system parameter which is influenced by material, hardness, roughness, wheel diameter, temperature, humidity, and aging processes [2-3].	Rolling friction appears between the wheels of a floor conveyor and the floor as well as between movable components (*e.g.*, lifting frame, rolls of continuous conveyors).
Approaches to analyze initial situation	Implementation
Measuring friction coefficients is possible with specific test facilities [2].	A lower friction reduces the required drive power. Therefore, it is important to consider the appropriate friction resistances when dimensioning a drive.
Benchmark	
As a rough approximation, the rolling resistance coefficient may be assumed as 1 to 2 % [1].	

Benefit and effort		
Expected benefit	The direct benefit initially influences the friction coefficients. Since the friction coefficient is an important parameter to dimension a drive, this leads to lower energy consumption. When existing drives are used further on, the savings effect depends on the drive characteristics.	
Side effects	Low friction losses reduce the wear of the components, which in turn decreases maintenance costs [5].	

Background	**External information sources**
Terms, definitions and theoretical explanations	Standardization
The *rolling resistance* F_{WR} is calculated by multiplying the rolling resistance coefficient w_R with the mass m and the gravity g [1].	The dimensioning of drives is supported through leaflets of the manufacturers, for example as provided in [6].

[1]   Martin, H.; Römisch, P.; Weidlich, A. (2008). Materialflusstechnik – Auswahl und Berechnung von Elementen und Baugruppen der Fördertechnik. 9th ed. Wiesbaden: Vieweg.
[2]   Nendel, K.; Mitzschke, F. (2006). Kunststoffgleitpaarungen in der Fördertechnik – Methoden zur Messung von Reibungswerten. In: Logistics Journal.
[3]   Jodin, D.; ten Hompel, M. (2012). Sortier- und Verteilsysteme – Grundlagen, Aufbau, Berechnung und Realisierung. 2nd ed. Berlin, Heidelberg: Springer.
[4]   Günthner, W. A.; Galka, S.; Tenerowicz, P. (2009). Roadmap für eine nachhaltige Intralogistik. In: Tagungsband zur 14. Wissenschaftlichen Fachtagung „Sustainable Logistics". Magdeburg, pp. 205-219.
[5]   Hallo, S.; Sumpf, J.; Nendel, K.; Drechsler, F. (2011). Energieeffizienz – Kennzeichen zukünftiger Fördertechnik. In: Hebezeuge Fördermittel 51 (10), pp. 502-506.
[6]   SEW Eurodrive (2001). Praxis der Antriebstechnik: Antriebe projektieren. URL: http://download.sew-eurodrive.com/download/pdf/10522905.pdf (visited on September 30, 2016).

Number	T15	Date	XX.XX.XXXX

Figure A12: Energy efficiency measure implementation sheet for case study 2 – planning of logistics systems – Measure T15 "Reduce rolling resistance of logistics equipment"

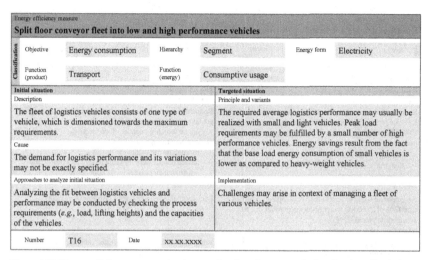

Figure A13: Energy efficiency measure implementation sheet for case study 2 – planning of logistics systems – Measure T16 "Split conveyor fleet into low and high performance vehicles"

Energy efficiency measure						
Reduce number of acceleration processes						

	Objective	Energy consumption	Hierarchy	Work center	Energy form	Electricity
Classification	Function (product)	Transport	Function (energy)	Consumptive usage		

Initial situation	**Targeted situation**
Description	Principle and variants
Stopping and accelerating a transport vehicle is part of any logistics process but the number of acceleration processes should be reduced.	Reducing the number of acceleration processes can be achieved by, among others [2]:
Cause	– layout with low number of intersections,
– waiting of vehicles – high traffic volume – unnecessary movements	– one-way transport system, – calming traffic within the factory, – reduce unnecessary movements (*e.g.*, shunting), and – sensitizing employees using floor conveyors.
Relevance	Application area
An empirical study in an automotive company identified that 47 % of a forklift's energy consumption is due to acceleration processes [1].	Focus on heavy-weight and/or fast-driving vehicles
Approaches to analyze initial situation	Employee involvement
– analyze operational states of a vehicle regarding time and energy consumption – document stopping and waiting times	When applied at manually operated vehicles, trainings for employees are recommended.

Benefit and effort	Side effects	– longer utilization time of batteries in logistics systems – calm traffic increases production safety

Background	**External information sources**
Terms, definitions and theoretical explanations	Standardization
The energy demand for accelerating depends on mass (transport system and goods) and targeted velocity [3].	Data sheets of forklift trucks according to VDI 2198 contain acceleration time and energy consumption within a standardized cycle (including 60 acceleration processes) [4].
Theoretical energy consumption	
The physical work to accelerate 1 ton to a velocity of 10 km/h accounts for: $W = \frac{1}{2}mv^2 \sim 1\,Wh$	

[1] Veit, T.; Fischer, S.; Strauch, J.; Krause, A. (2010). Umsetzung logistischer Strategien unter Berücksichtigung der Energieeffizienz. In: Tagungsband zum 4. Symposium Wissenschaft und Praxis & 8. Fachtagung Vernetzt Planen und Produzieren – VPP 2010, Chemnitz.
[2] Krones, M.; Hopf, H.; Müller, E. (2014). Ermittlung von Energieeffizienzmaßnahmen für Planung und Betrieb von Logistiksystemen. In: Proceedings of the 3rd International Chemnitz Manufacturing Colloquium 2014 – Proceedings Part 2. Chemnitz, pp. 499-509.
[3] Martin, H. (2011). Transport- und Lagerlogistik – Planung, Struktur, Steuerung und Kosten von Systemen der Intralogistik. 8th ed. Wiesbaden: Vieweg+Teubner.
[4] Verein Deutscher Ingenieure (2002). VDI-Guideline 2198 Type Sheets for Industrial Trucks. Berlin: Beuth.

Number	T18	Date	XX.XX.XXXX

Figure A14: Energy efficiency measure implementation sheet for case study 2 – planning of logistics systems – Measure T18 "Reduce number of acceleration processes"

Energy efficiency measure						
Reduce mass of containers and packaging material						
Classification	Objective	Energy consumption	Hierarchy	Work center	Energy form	Electricity, compressed air
	Function (product)	Transport, storage	Function (energy)	Consumptive usage		

Initial situation	Targeted situation
Description	Principle and variants
Since the mass of goods may usually not be influenced with regard to energy efficiency, additional masses of containers and packaging material might be reduced.	Reducing the mass may be achieved by: – container design, – lightweight components, and – reducing packaging.
Approaches to analyze initial situation	
A helpful indicator is the ratio between own mass and useful load capacity. This ratio should be minimized.	
Benchmark	Application area
The share of energy consumption to move a conveyor without load is approx. 80 % for a forklift truck and more than 90 % for a storage and retrieval machine [1].	The mass needs to be reduced especially for dynamic movements [2]. Hence, the measure is important for uncontinuous conveyors, which are characterized by frequent changes in the direction of the movement [1].

Benefit and effort

Expected benefit	The theoretical benefit can be calculated by the effect on the kinetic energy: $$E = \frac{1}{2}mv^2$$ Therefore, a reduction of the mass directly leads to the same proportion of energy savings.
Side effects	A higher utilization rate increases the transport mass and may conflict with the EEM „Reduce masses to be transported". The trade-off needs to be analyzed individually.

Sources

[1] Furmans, K.; Linsel, P.(2011). Leichtbau bei Unstetigförderern durch Einsatz moderner Werkstoffe. In: Logistics Journal.
[2] Hülsmann, S.; Köpschall, M.; Neumann, R.; Ohmer, M.; Hobusch, G.; Ruppelt, E.; Doll, M.; Krichel, S.; Sawodny, O.; Elsland, R.; Hirzel, S.; Schröter, M.; Weißfloch, U.; Blank, F.; Nguyen, Q. K.; Roth-Stetlow, J. (2012). EnEffAH – Energieeffizienz in der Produktion im Bereich Antriebs- und Handhabungstechnik. URL: http://www.eneffah.de/EnEffAH_Broschuere.pdf (visited on September 30, 2016).

Number	T19	Date	XX.XX.XXXX

Figure A15: Energy efficiency measure implementation sheet for case study 2 – planning of logistics systems – Measure T19 "Reduce mass of containers and packaging material"

Energy efficiency measure				
Switch equipment off or into stand-by mode appropriately				

Classification				
Objective	Energy consumption	Hierarchy	Work center	Energy form	Electricity
Function (product)	Transport, storage, information and communication technology		Function (energy)	Consumptive usage

Initial situation	Targeted situation
Description	Principle and variants
Some equipment may be running continuously although this is not required during waiting time (*e.g.*, sorting facilities). Besides logistics equipment, this measure includes other general facilities (*e.g.*, computers).	The purpose is to operate equipment only during times, at which it is necessary; during other times, it should be switched off or in stand-by mode. This may be achieved by program control or by sensitizing employees.
	Application area
	Switching off equipment is most recommended, where there is a high difference in energy consumption between off/ stand-by and operating. Furthermore, possible challenges with frequent switching processes need to be considered.
	Employee involvement
	When employees need to switch off manually, they should be informed about what equipment should be switched off when and how (*i.e.*, detailed work instruction) and reminded by using easy visual leaflets.

Benefit and effort		
Expected benefit	If controls of conveyor, sorter, and storage equipment are switched off strictly, 5 to 15 % of the energy consumption may be saved [1]. One advantage of the measure is the usually low investment.	

Sources		
[1]	Kramer, J. (2012). Energiesparpotenziale im Lager. In: Logistra, 24 (1-2), pp. 17-19.	

Number	T20	Date	XX.XX.XXXX

Figure A16: Energy efficiency measure implementation sheet for case study 2 – planning of logistics systems – Measure T20 "Switch equipment off or into stand-by mode appropriately"

Energy efficiency measure					
Summarize transport orders					

Classification	Objective	Energy consumption	Hierarchy	Work center, segment	Energy form	Electricity
	Function (product)	Transport	Function (energy)	Consumptive usage		

Initial situation	**Targeted situation**
Description	Principle and variants
Realizing transport orders with separate transport routes reduces delivery time but increases the number of transports and, thereby, reduces the energy consumption.	A more efficient concept is to summarize transport orders in order to increase the utilization rate of logistics equipment. This reduces the number of transport processes while increasing the transported mass. The effect on the energy consumption needs to be assessed for the individual case. An example for summarizing transport orders is the substitution of forklift trucks by electric tractors [1]. In this case, the traffic is harmonized additionally. Another example is to use double cycles instead of single cycles for storage and retrieval machines, which reduces the average energy consumption per load unit [2].
	Application area
	Summarizing transport orders is especially recommended, if the additional transport mass does hardly influence the energy consumption, *i.e.*, for logistics vehicles with a relatively high base load consumption.

Benefit and effort

Side effects — A higher utilization rate increases the transport mass and may conflict with the EEM „Reduce masses to be transported". The trade-off needs to be analyzed individually. Summarizing transport orders may be supported by a fleet management (see EEM „Use vehicle fleet management").

Background

Terms, definitions and theoretical explanations

The *energy efficiency* of a logistics process may be represented by the ratio of the energy consumption and the logistics performance (*e.g.*, product of mass and transport distance).

Sources

[1] Droste, M.; Hasselmann, V.-R.; Deuse, J. (2012). Optimierung innerbetrieblicher Milkrun-Systeme – Entwicklung eines parameterbasierten Modells zur Optimierung der Materialbereitstellung. In: Productivity Management, 17 (1), pp. 25-27.

[2] Siegel, A.; Schulz, R.; Turek, K.; Schmidt, T.; Zadek, H. (2013). Modellierung des Energiebedarfs von Regalbediengeräten und verschiedener Lagerbetriebsstrategien zur Reduzierung des Energiebedarfs. In: Logistics Journal.

Number	T21	Date	XX.XX.XXXX

Figure A17: Energy efficiency measure implementation sheet for case study 2 – planning of logistics systems – Measure T21 "Summarize transport orders"

Energy efficiency measure					
Reduce unnecessary movements of logistics vehicles					

Classification	Objective	Energy consumption	Hierarchy	Work center	Energy form	Electricity
	Function (product)	Transport	Function (energy)	Consumptive usage		

Initial situation	**Targeted situation**
Description	Principle and variants
Each transport process causes energy consumption, whereof some processes do not add value. These movements should be avoided.	Shunt movements may be reduced by a better transport organization (*e.g.*, time schedule), harmonized logistics processes or manually movable transport containers. Unnecessary empty trips may be reduced by using a fleet management system, which optimizes the assignment of transport orders.
Cause	Employee involvement
There are several causes for unnecessary movements in logistics processes. Shunt movements may be necessary for sorting transport containers either in the warehouse or at the demanded position (*e.g.*, assembly line). A similar aspect are turning manoeuvers, for example in dead end aisles, or tours to check the amount of available material.	The measure gains from the sensitization of employees (*e.g.*, forklift drivers). This includes qualifications, trainings, and visual support.

Benefit and effort	Expected benefit	The energy savings by reducing unnecessary movements result from lower transport distances. Furthermore, dynamic movements (such as shunting) usually have a high frequency of acceleration processes, which are especially energy-intensive.
	Number	T22	Date	XX.XX.XXXX

Figure A18: Energy efficiency measure implementation sheet for case study 2 – planning of logistics systems – Measure T22 "Reduce unnecessary movements of logistics vehicles"

Energy efficiency measure
Reduce driving velocity

	Objective	Energy consumption	Hierarchy	Work center	Energy form	Electricity
Classification	Function (product)	Transport	Function (energy)	Consumptive usage		

Initial situation	**Targeted situation**
Description	Principle and variants
Usually, equipment is adjusted to perform a logistics task in minimal time. This may lead to high velocities and dynamic movements in logistics processes, which increase the energy consumption.	Reducing the driving velocity has two effects [1]: First, the acceleration time decreases since the maximum velocity is achieved earlier. This reduces acceleration energy consumption. The second aspect is the lower energy consumption during movement with constant velocity due to lower movement resistance. On the other hand, the transport time increases. The optimum needs to be found individually.
Approaches to analyze initial situation	Employee involvement
Potentials to reduce the driving velocity may be present when logistics vehicles have a high share of waiting time.	The velocity of manually operated vehicles is mainly affected by the operators, which need information and trainings towards their influence.

	Expected benefit	The energy savings may vary greatly. An example of a continuous conveyor system showed a saving potential of up to 40 % when reducing the velocity [2].
Benefit and effort	Side effects	A reduced driving velocity may lead to decreasing productivity.

Background	**External information sources**
Theoretical energy consumption	Standardization
The energy that is theoretically required to bring a 1,000 kg vehicle to a velocity of 12 km/h is 5.5 kWs.	The velocity of conveyors is usually part of the data sheet as provided from the manufacturer.

Sources	[1]	Schulz, R.; Monecke, J.; Zadek, H. (2012). Der Einfluss kinematischer Parameter auf den Energiebedarf eines Regalbediengerätes. In: Logistics Journal.
	[2]	Müller, E.; Krones, M.; Strauch, J.; Fischer, S.; Veit, M. (2013). Energieeffizienz von Stetigförderern – Analyse und Bewertung von Maßnahmen in Planung und Betrieb. In: Tagungsband 18. Magdeburger Logistiktage. Magdeburg, pp. 153-160.

Number	T23	Date	XX.XX.XXXX

Figure A19: Energy efficiency measure implementation sheet for case study 2 – planning of logistics systems – Measure T23 "Reduce driving velocity"

Energy efficiency measure						
Arrange warehouse racks in order to minimize transport routes						
Objective	Energy consumption		Hierarchy	Work center, segment, division, building		
Energy form	Electricity		Function (product)	Transport, storage	Function (energy)	Consumptive usage

Initial situation	**Targeted situation**
Description	Principle and variants
The warehouse racks have a storage strategy (*i.e.*, dynamic or static), which is not yet adjusted to the requirements of energy efficiency.	The layout of the warehouse should be optimized in terms of the lowest combination of transport frequency, transport distance and transport mass, *i.e.*, reducing the energy consumption is achieved by transporting heavy material as short as possible. The focus on the transport frequency is usually referred to as 'ABC classification'.
Approaches to analyze initial situation	Application area
An analysis contains documenting the transport frequencies for each warehouse rack, assigned to the respective material units.	The measure is applicable to various warehouse structures. Besides the layout of the racks, this aspect is especially important for the vertical positions in a high rack warehouse [1].

Sources			
[1] Braun, M.; Schönung, F.; Furmans, K. (2012). Energieeffizienz beim Lager- und Kommissioniervorgang. In: Productivity Management, 17 (4), pp. 29-32.			
Number	T24	Date	XX.XX.XXXX

Figure A20: Energy efficiency measure implementation sheet for case study 2 – planning of logistics systems – Measure T24 "Arrange warehouse racks in order to minimize transport routes"

Energy efficiency measure					
Reduce areas for transport and providing material					

Classification					
Objective	Energy consumption	Hierarchy	Work center, segment, division, building		
Energy form	Electricity, Heat	Function (product)	Transport, building services	Function (energy)	Consumptive usage

Initial situation	Targeted situation
Description	Principle and variants
The energy consumption for the building services of a warehouse (*e.g.*, heating, lighting) is influenced by the required area.	The storage area may be reduced by concentrated storage (*e.g.*, high rack warehouses, small overpack).
Cause	The transport area may be reduced by, among others:
Areas in a warehouse are required for storage, transport paths and workplaces (*e.g.*, order-picking).	– using small vehicles, – applying one-way traffic, – reducing the turning radius of vehicles (*e.g.*, number of
Approaches to analyze initial situation	trailers), and
The warehouse layout needs to be analyzed while distinguishing between the purpose of the areas.	– reducing the width of transport paths (*e.g.*, automatic areas have lower requirements for safety distances).

Benefit and effort		
Expected benefit	An average warehouse consumes 38 to 108 kWh/m^2 heating energy, 11 to 13 kWh/m^2 lighting energy and 17 to 20 kWh/m^2 electricity for logistics equipment [1, 2].	

Sources		
[1] Rinza, T.: Effizienter Materialfluss. In: Automobil Industrie 55 (2010) 4, S. 46-47. [2] Kramer, J.: Energiesparpotenziale im Lager. In: Logistra 24 (2012) 1-2, S. 17-19.		

Number	T25	Date	XX.XX.XXXX

Figure A21: Energy efficiency measure implementation sheet for case study 2 – planning of logistics systems – Measure T25 "Reduce areas for transport and providing material"

Energy efficiency measure					
Regularly carry out a compensation charge of batteries					

Classification					
Objective	Energy consumption	Hierarchy	Work center	Energy form	Electricity
Function (product)	Transport	Function (energy)	Conversion		

Initial situation	Targeted situation
Description	Principle and variants
Due to variations in the performance of battery cells, the state of charging may vary between the cells.	A compensation charge compensates the charge differences between the battery cells. Hence, it avoids cell defects and increases the service life of the battery.
	Application area
	The manual compensation charge refers to lead acid batteries with liquid electrolyt.
	Implementation
	The trigger for a compensation charge may be either time (*e.g.*, once a week) or number of charging/discharging processes. Furthermore, saisonally used batteries demand a compensation charge before or after the season. Details on the implementation can be found in the data sheet of a battery.

Number	T127	Date	XX.XX.XXXX

Figure A22: Energy efficiency measure implementation sheet for case study 2 – planning of logistics systems – Measure T127 "Regularly carry out a compensation charge of batteries"

Energy efficiency measure						
Reduce room height						
Classification	Objective	Energy consumption	Hierarchy	Building	Energy form	Electricity, Heat
	Function (product)	Building structure, heating, cooling, air-conditioning			Function (energy)	Consumptive usage

Initial situation	**Targeted situation**
Description	Principle and variants
The volume of a room influences the required heating power. Furthermore, warm air rises under the roof, where it is not necessary and where heat losses may be higher due to worse insulation at the roof (as compared to walls) [1].	The room height should be adjusted to the requirements and processes in the building.

Benefit and effort

Side effects   A reduced room height allows the realization of other measures (see EEM „Reduce height of lamps").

Background	**External information sources**
Theoretical energy consumption	Standardization
The energy that is theoretically required to heat a volume is calculated according to $$Q = V \cdot s \cdot \Delta T$$ with volume V, material-specific heat accumulating capacity s and temperature difference ΔT. Hence, heating up a building with an area of 10,000 m² and a height of 14 m from 15 °C to 19 °C theoretically requires 190 kWh.	The energy demand for heating or cooling a building may be determined according to DIN V 18599-2 [2].

Sources

[1] EnergieAgentur Nordrhein-Westfalen (2010). Heizung – Potenziale zur Energieeinsparung. URL: https://www2.duesseldorf.de/fileadmin/Amt19/saga/doc/pdf/8_bf_heizung_ea_nrw.pdf (visited on September 30, 2016).
[2] Deutsches Institut für Normung (2011). DIN V 18599 Energy Efficiency of Buildings – Calculation of the Net, Final and Primary Energy Demand for Heating, Cooling, Ventilation, Domestic Hot Water and Lighting – Part 2: Net Energy Demand for Heating and Cooling of Building Zones. Berlin: Beuth.

Number	H29	Date	XX.XX.XXXX

Figure A23: Energy efficiency measure implementation sheet for case study 2 – planning of logistics systems – Measure H29 "Reduce room height"

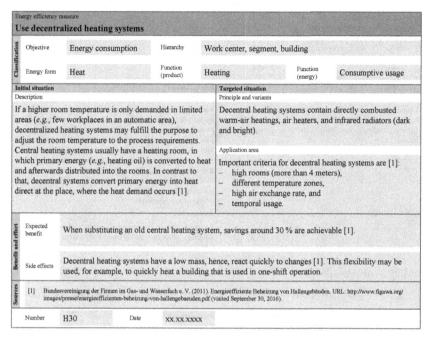

Energy efficiency measure					
Use decentralized heating systems					
Objective	Energy consumption	Hierarchy	Work center, segment, building		
Energy form	Heat	Function (product)	Heating	Function (energy)	Consumptive usage

Initial situation	Targeted situation
Description	Principle and variants
If a higher room temperature is only demanded in limited areas (*e.g.*, few workplaces in an automatic area), decentralized heating systems may fulfill the purpose to adjust the room temperature to the process requirements. Central heating systems usually have a heating room, in which primary energy (*e.g.*, heating oil) is converted to heat and afterwards distributed into the rooms. In contrast to that, decentral systems convert primary energy into heat direct at the place, where the heat demand occurs [1].	Decentral heating systems contain directly combusted warm-air heatings, air heaters, and infrared radiators (dark and bright).
	Application area
	Important criteria for decentral heating systems are [1]: – high rooms (more than 4 meters), – different temperature zones, – high air exchange rate, and – temporal usage.

Expected benefit	When substituting an old central heating system, savings around 30 % are achievable [1].
Side effects	Decentral heating systems have a low mass, hence, react quickly to changes [1]. This flexibility may be used, for example, to quickly heat a building that is used in one-shift operation.

[1]	Bundesvereinigung der Firmen im Gas- und Wasserfach e. V. (2011). Energieeffiziente Beheizung von Hallengebäuden. URL: http://www.figawa.org/images/presse/energieeffizienten-beheizung-von-hallengebaeuden.pdf (visited September 30, 2016).

Number	H30	Date	XX.XX.XXXX

Figure A24: Energy efficiency measure implementation sheet for case study 2 – planning of logistics systems – Measure H30 "Use decentralized heating systems"

Energy efficiency measure						
Use radiating heating systems						

	Objective	Energy consumption	Hierarchy	Segment, Building	Energy form	Heat
Classification	Function (product)	Heating	Function (energy)	Consumptive usage		

Initial situation	**Targeted situation**
Description	Principle and variants
Usually, warm air heating systems are applied in industrial hall buildings. However, depending on the requirements, radiating heating systems may be the more energy-efficient choice.	Radiating heating systems may either be designed as direct systems (*e.g.*, dark radiator) or as surface heating (*e.g.*, floor heating). The surface heating has the advantage of lower flow temperatures. Furthermore, the heat is not transferred into the air but to the surrounding surfaces almost free of any loss [1]. This reduces ventilation heat losses (*e.g.*, through open gates). By using radiating heating systems, the room temperature may be reduced by 3 to 5 °C while achieving the same thermal comfort [2] – see EEM „Reduce average room temperature".
	Application area
	Radiating heating systems are recommended for high buildings (since they hardly cause temperature layering) and buildings with frequent door openings [1].

	Energy savings	up to 30 % [2]
Benefit and effort	Expected benefit	Radiating heatings may save between 25 and 30 % energy as compared to warm-air heating systems [3].
	Side effects	Other positive side effects of using a radiating heating system are [3-4]: – avoid resuspension of dust-laden air, – specifically heat single workplaces, – avoid recirculation of large air masses (see EEM „Reduce heat losses through in-bound and out-bound transport processes"), – short pre-heating time, and – avoid draft effects. Due to lower flow temperatures, radiating surface heating systems may be more easily combined with regenerative energies (*e.g.*, ground water, biomass) [5]. On the other hand, negative side effects are [6]: – functionality reduced to heating (*i.e.*, fresh air needs to be provided separately), – surface temperature of products may be too high (*e.g.*, pharmaceutical industry), and – limited applicability in rooms with explosive gases or vapors.

	[1]	Energieagentur Nordrhein-Westfalen (2007). Beheizung von Hallen und hohen Räumen. URL: http://www.energieagentur.nrw/content/anlagen/Hallenheizung04062007.pdf (visited on September 30, 2016).
Sources	[2]	Kramer, J. (2012). Energiesparpotenziale im Lager. In: Logistra 24 (1-2), pp. 17-19.
	[3]	Günthner, W. A.; Galka, S.; Tenerowicz, P. (2009). Roadmap für eine nachhaltige Intralogistik. In: Tagungsband zur 14. Wissenschaftlichen Fachtagung „Sustainable Logistics". Magdeburg, pp. 205-219.
	[4]	Kramer, J. (2012). Energieeinsparpotenziale rund um das Flurförderzeug. In: Hebezeuge Fördermittel 52 (4), pp. 186-188.
	[5]	Günthner, W. A.; Hausladen, G.; Freis, J.; Vohlidka, P. (2014). Das CO₂-neutrale Logistikzentrum – Entwicklung von Handlungsempfehlungen für energieeffiziente Logistikzentren. URL: http://www.fml.mw.tum.de/fml/index.php?Set_ID=870&Download=Forschungsbericht_Das_CO2_neutrale_Logistikzentrum_IGF_398ZN (visited on September 30, 2016).
	[6]	BDEW Bundesverband der Energie- und Wasserwirtschaft e. V. (2010). Strahlungsheizung Erdgas-Infrarotheizsysteme. URL: http://www.gewerbegas-online.de/fileadmin/user_upload/dokumente/Erdgas-Infrarotheizsysteme_8.pdf (visited on September 30, 2016).

Number	H81	Date	XX.XX.XXXX

Figure A25: Energy efficiency measure implementation sheet for case study 2 – planning of logistics systems – Measure H81 "Use radiating heating systems"

Energy efficiency measure				
Reduce average room temperature				

Classification

Objective	Energy consumption	Hierarchy	Segment	Energy form	Heat

Function (product)	Heating	Function (energy)	Consumptive usage

Initial situation	**Targeted situation**
Description	Principle and variants

The room temperature may be too high for the requirements of products, processes, and personnel due to changes in the production process which have not been passed on to the heating system.	The perceived room temperature t_R is influenced by the air temperature t_A and the radiation intensity I_S of the surrounding surfaces [2]:

$$t_R = t_A + k \cdot I_S,$$

whereof the factor k depends on the configuration of the radiating heating system as described in the following table.

Installation of radiating heating system	Factor k
Radiation from one side (*e.g.*, workplace heating)	0.025
Raidation from two sides with vertical radiators	0.033
Radiation from two sides with diagonal radiators	0.045
Radiation from four sides with diagonal radiators	0.072

Approaches to analyze initial situation	Employee involvement
The air temperature is measured 60 cm above the floor for seated working activities and 1.1 m above the floor for upright activities [1].	Due to possible effects on the personnel, the works council should be considered before realizing this measure.

Benefit and effort

Expected benefit	As a rough orientation, reducing the room temperature by one degree saves 6 % of heating energy demand [3].

Background	**External information sources**
Terms, definitions and theoretical explanations	Standardization
The *room temperature* is the temperature as perceived by the staff. The *air temperature* is the temperature of the surrounding air without influence of heat radiation.	Details on temperature measurements are defined in DIN EN ISO 7726 [4].
	Legislation
	The workplace regulation ASR A3.5 defines the following room temperatures depending on the working conditions [1]:

Predominant body posture	Work intensity		
	light	medium	heavy
Seated	20 °C	19 °C	-
Upright or walking	19 °C	17 °C	12 °C

Sources

[1] Ausschuss für Arbeitsstätten, ASTA (2010). Arbeitsstätten-Richtlinie ASR A3.5 Raumtemperatur.
[2] BDEW Bundesverband der Energie- und Wasserwirtschaft e. V. (2010). Strahlungsheizung Erdgas-Infraroheizsysteme. URL: http://www.gewerbegas-online.de/fileadmin/user_upload/dokumente/Erdgas-Infrarotheizsysteme_8.pdf (visited on September 30, 2016).
[3] Energieagentur Nordrhein-Westfalen (2010). Heizung – Potenziale zur Energieeinsparung. URL: https://www2.duesseldorf.de/fileadmin/Amt19/saga/doc/pdf/8_bf_heizung_ea_nrw.pdf (visited on September 30, 2016).
[4] Deutsches Institut für Normung (2002). DIN EN ISO 7726 Ergonomics of the Thermal Environment – Instruments for Measuring Physical Quantities. Berlin: Beuth.

Number	H92	Date	XX.XX.XXXX

Figure A26: Energy efficiency measure implementation sheet for case study 2 – planning of logistics systems – Measure H92 "Reduce average room temperature"

Energy efficiency measure					
Reduce ratio between surface and volume of a building					

Classification	Objective	Energy consumption	Hierarchy	Building	Energy form	Heat
	Function (product)	Building structure	Function (energy)	Consumptive usage		

Initial situation	**Targeted situation**
Description	Principle and variants
The ratio between surface and volume of a building, the cubage, is an important parameter when designing a new building.	The form and size of a building influences the enveloping surface. For example, U-shaped buildings should be avoided. Instead, favorable elements include:
	– arrange usable space on both sides of routes and paths
Cause	– large buildings instead of several small buildings
The transmission heat losses of a building are influenced by the enveloping surface. Hence, a high surface in relation to the building volume increases transmission losses.	– quadratic base area
	Application area
Benchmark	Besides the surface area, transmission heat losses are affected by the difference between inside and outside temperatures. Hence, the measure is especially recommended for buildings and areas with a high temperature difference.
For large buildings, a ratio as low as 0.2 m^2/m^3 can be achieved [1].	

Benefit and effort	Expected benefit	Reducing the ratio between surface area and volume by 0.1 m^3/m^3 reduces the heating energy consumption by approximately 10 kWh/m^2 per year [2]. A study on the energy consumption of logistics buildings analyzed the effect of the ratio between surface and volume: As compared to a quadratic building, a doubled length and halved height led to an increase in the heating energy consumption of 20 % [3].

Background
Terms, definitions and theoretical explanations
Transmission heat losses occur through closed surfaces (heat conduction) due to a temperature difference.

Sources	[1]	Müller, E.; Engelmann, J.; Löffler, T.; Strauch, J. (2009). Energieeffiziente Fabriken planen und betreiben. Berlin, Heidelberg: Springer.
	[2]	Dilmetz, K.; Erhorn, H. (2004). Bedeutung der Energieeinsparung im Gebäudebereich. In: Joos, L.: Energieeinsparung in Gebäuden – Stand der Technik und Entwicklungstendenzen. 2nd ed. Essen: Vulkan, pp. 1-62.
	[3]	Günthner, W. A.; Hausladen, G.; Freis, J.; Vohlidka, P. (2014). Das CO₂-neutrale Logistikzentrum – Entwicklung von Handlungsempfehlungen für energieeffiziente Logistikzentren. URL: http://www.fml.mw.tum.de/fml/index.php?Set_ID=870&Download=Forschungsbericht_Das_CO2_neutrale_Logistikzentrum_IGF_398ZN (visited on September 30, 2016).

Number	H169	Date	XX.XX.XXXX

Figure A27: Energy efficiency measure implementation sheet for case study 2 – planning of logistics systems – Measure H169 "Reduce ratio between surface and volume of a building"

Energy efficiency measure					
Adjust illumination to working task					

Classification	Objective	Energy consumption	Hierarchy	Work center, segment	Energy form	Electricity
	Function (product)	Lighting	Function (energy)	Consumptive usage		

Initial situation	**Targeted situation**
Description	Principle and variants

When planning a new or checking an existing lighting system, the illumination needs to be adjusted to the work task.

The minimum lighting requirements are defined in norm DIN EN 12464 [4] and workplace regulation ASR 3.4 [1], for example:

Approaches to analyze initial situation

Measuring the illumination should be conducted at a height of 0.75 m above the floor for seated activities and 0.85 m for upright activities [1]. The grid of measuring points may be selected depending on the size of the room [2]: 0.6 m for small rooms, 1 m for medium-sized rooms and 3 m for large rooms. Furthermore, the measurement in the task area should be performed every 0.2 m.

Standard use	Minimum illumination in lx
Circulation area	50-100
Storage	200
Sanitary facilities	200
Metal-working processes	200-300
Assembly tasks	200-750
Measurement tasks	500
Offices	500
Surface processing	750
Tool manufacturing	1,000
Inspection	1,000

Benchmark

The VDI guideline 3807 describes benchmark values of the lighting energy consumption for various usage scenarios [3], for example:

Standard use	Specific lighting energy demand in kWh/m² · a				
	very high	high	average	low	very low
Circulation area	37	27	12	6	1
Storage	38	34	18	12	0,6
Assembly & manufacture	80	66	36	18	10

Background
Terms, definitions and theoretical explanations

The *luminous flux* (measured in lm) is the total lighting energy, which is emitted from a source into all directions. The *illumination* is the ratio between the luminous flux and the area or the surface, to which it is emitted. The illumination is measured in lx with $1\,lx = 1\,lm/m^2$. [5]

[1] Ausschuss für Arbeitsstätten, ASTA (2011). Arbeitsstätten-Richtlinie ASR A3.4 Beleuchtung.
[2] Fördergemeinschaft Gutes Licht; LiTG Deutsche Lichttechnische Gesellschaft e. V. (2013). Leitfaden zur DIN EN 12464-1: Beleuchtung von Arbeitsstätten – Teil 1: Arbeitsstätten in Innenräumen. URL: https://www.trilux.com/fileadmin/Downloads/Leitfaden_DIN_2.Auflage_Lichtwissen.pdf (visited on September 30, 2016).
[3] Verein Deutscher Ingenieure (2008). VDI-Guideline 3807 Characteristic Values for Energy and Water Consumption of Buildings – Part 4: Characteristic Values for Electrical Energy. Berlin: Beuth.
[4] Deutsches Institut für Normung (2011). DIN EN 12464 Light and Lighting – Part 1: Lighting of Workplace. Berlin: Beuth.
[5] Hesselbach, J. (2012). Energie- und klimaeffiziente Produktion – Grundlagen, Leitlinien und Praxisbeispiele. Wiesbaden: Vieweg+Teubner.

Number	L26	Date	XX.XX.XXXX

Figure A28: Energy efficiency measure implementation sheet for case study 2 – planning of logistics systems – Measure L26 "Adjust illumination to working task"

Energy efficiency measure						
Split lighting into general and workspace lighting						
Objective	Energy consumption	Hierarchy	Work center, segment	Energy form	Electricity	
Function (product)	Lighting	Function (energy)	Consumptive usage			

Initial situation	**Targeted situation**
Cause	Principle and variants
A general lighting is advantageous when workspaces cannot be spatially assigned during planning or when a flexible layout is desired [1].	The energy saving effect results from the fact that the general lighting is dimensioned with a lower illumination. The workspace lighting provides the higher illumination only at places, where it is required.
Approaches to analyze initial situation	Application area
It is necessary to define spaces and their lighting requirements, which depend on the usage scenario of the respective area.	The use of specific workspace lighting makes sense for areas with varying lighting requirements.
	Implementation
	The surrounding area around the workspace (at least 0.5 m to each side) needs to fulfill the minimum illumination requirements as described in DIN 12464 [2] and workplace regulation A3.4 [1].

Number	L27	Date	XX.XX.XXXX

Figure A29: Energy efficiency measure implementation sheet for case study 2 – planning of logistics systems – Measure L27 "Split lighting into general and workspace lighting"

Energy efficiency measure						
Reduce height of lamps						
Classification	Objective	Energy consumption	Hierarchy	Work center, segment, building		
	Energy form	Electricity	Function (product)	Lighting	Function (energy)	Consumptive usage

Initial situation	Targeted situation
Description	Principle and variants
Since the illuminous flux is emitted in all directions from a light source, the illumination at the floor (or workspace) is reduced with an increasing height of the lamps.	If the ceiling height is not entirely used for working tasks, the lamps should be placed as low as possible. This reduces the number of required lamps. [1]

Benefit and effort

Expected benefit: Reducing the height of lamps may have a savings potential of up to 20 % when the rated lighting power is adjusted accordingly [1]. As a rough orientation, an earlier working regulation on artifical lighting describes the effect of an increasing lamp height on lighting power: According to this, a height of 4 m increases the power demand by 28 % as compared to a height of 2 m [2].

Background

Terms, definitions and theoretical explanations

The *luminous flux* (measured in lm) is the total lighting energy, which is emitted from a source into all directions. The *illumination* is the ratio between the luminous flux and the area or the surface, to which it is emitted. The illumination is measured in lx with $1\,lx = 1\,lm/m^2$. [3]

[1] EnergieAgentur Nordrhein-Westfalen (2010). Beleuchtung – Potenziale zur Energieeinsparung. URL: https://services.nordrheinwestfalendirekt.de/broschuerenservice/download/70588/qb_beleuchtung_final.pdf (visited September 30, 2016).
[2] Ausschuss für Arbeitsstätten, ASTA (1993). Arbeitsstätten-Richtlinie ASR A7.3 Künstliche Beleuchtung.
[3] Hesselbach, J. (2012). Energie- und klimaeffiziente Produktion – Grundlagen, Leitlinien und Praxisbeispiele. Wiesbaden: Vieweg+Teubner.

Number	L73	Date	XX.XX.XXXX

Figure A30: Energy efficiency measure implementation sheet for case study 2 – planning of logistics systems – Measure L73 "Reduce height of lamps"

Energy efficiency measure						
Place windows in order to maximize daylight use						
Classification	Objective	Energy consumption	Hierarchy	Segment, division, building		
	Energy form	Electricity	Function (product)	Building structure, lighting	Function (energy)	Consumptive usage

Initial situation	**Targeted situation**
Description	Principle and variants
The use of daylight should be maximized since daylight is the most efficient lighting and achieves the highest illumination [1].	The use of daylight may be achieved by windows in the walls or the ceiling (*e.g.*, shed roofs, skylights). Windows in walls should be placed up until the ceiling in order to increase the illumination.
	Application area
	Windows should be placed on the south side of a building in order to maximize the use of daylight. Windows in refrigerated warehouses should be minimized [2].
	Implementation
	Windows have a higher heat transition coefficient than walls. Hence, attention needs to be paid to the insulation standard of the window (*e.g.*, triple glazing). A solar protection is helpful in order to limit glare and overheating in summer. Daylight is used efficiently when combined with a lighting management (*i.e.*, dim or switch off light in dependence of daylight).
	Employee involvement
	The comfort and performance of employees is typically higher with daylight use instead of artifical light [2].
	Information need
	The share of useful daylight depends on the geometry of the rooms (*i.e.*, consider the room depth for windows and the room height for skylights).

Benefit and effort	Expected benefit	Dimming the lighting depending on daylight may save up to 20 % energy [3]. A consequent use of daylight may reduce the time, during which artificial light is required, below 50 % of the building usage time [4].
	Side effects	A positive side effect is higher comfort. On the other hand, windows have higher heat transmission losses than walls. Hence, the savings of lighting energy consumption need to be opposed to the increasing energy demand for heating [2].

Background	**External information sources**
Terms, definitions and theoretical explanations	Legislation
The *daylight factor* is the ratio between the illumination in a room and the illumination outside with overcast sky.	According to the workplace regulation A3.4, a workplace should have a daylight factor of at least 2 % (4 % when skylights are used). Furthermore, the ratio between translucent surfaces (*e.g.*, windows, skylights) and the floor area should be at least 10 %. [5]

Sources	
[1]	Hesselbach, J. (2012). Energie- und klimaeffiziente Produktion – Grundlagen, Leitlinien und Praxisbeispiele. Wiesbaden: Vieweg+Teubner.
[2]	Günther, W. A.; Hausladen, G.; Freis, J.; Vohlidka, P. (2014). Das CO₂-neutrale Logistikzentrum – Entwicklung von Handlungsempfehlungen für energieeffiziente Logistikzentren. URL: http://www.fml.mw.tum.de/fml/index.php?Set_ID=870&Download=Forschungsbericht_Das_CO2_neutrale_Logistikzentrum_IGF_398ZN (visited September 30, 2016).
[3]	ETAP Lighting (n.d.). Integrierte Lichtregelung. URL: http://www.etaplighting.com/uploadedFiles/Downloadable_documentation/documentatie/brochures_ETAP_verlichting/Integrierte%20Lichtregelung_DE.pdf (visited on September 30, 2016).
[4]	Bayerisches Landesamt für Umweltschutz (2004). Bürogebäude – Klima schützen – viel sparen mit wenig Strom. URL: http://www.lfu.bayern.de/energie/buerogebaude/leitfaden.pdf#page=23 (visited on September 30, 2016).
[5]	Ausschuss für Arbeitsstätten, ASTA (2011). Arbeitsstätten-Richtlinie ASR A3.4 Beleuchtung.

Number	L125	Date	XX.XX.XXXX

Figure A31: Energy efficiency measure implementation sheet for case study 2 – planning of logistics systems – Measure L125 "Place windows in order to maximize daylight use"

Printed in the United States
By Bookmasters